Musica Mathematica

Methodology of Music Research
Methodologie der Musikforschung
Edited by / Herausgegeben von Nico Schüler

Vol. / Bd. 9

PL ACADEMIC RESEARCH

Rima Povilionienė

Musica Mathematica

Traditions and Innovations in Contemporary Music

PL ACADEMIC RESEARCH

Bibliographic Information published by the Deutsche Nationalbibliothek
The Deutsche Nationalbibliothek lists this publication in the Deutsche Nationalbibliografie; detailed bibliographic data is available in the internet at http://dnb.d-nb.de.

Library of Congress Cataloging-in-Publication Data
Names: Povilionienė, Rima.
Title: Musica mathematica : traditions and innovations in contemporary music / Rima Povilionienė.
Other titles: Musica mathematica. English
Description: Frankfurt am Main ; New York : PL Academic Research, 2016. | Series: Methodology of music research ; Vol. 9 | Translated from Lithuanian. | Includes bibliographical references and index.
Identifiers: LCCN 2016048131 | ISBN 9783631713815
Subjects: LCSH: Music—Mathematics. | Music—Philosophy and aesthetics. | Symbolism of numbers in music.
Classification: LCC ML3800 .P81513 2016 | DDC 780/.051—dc23 LC record available at https://lccn.loc.gov/2016048131

This book was originally published in Lithuanian under the title *Musica mathematica. Tradicijos ir inovacijos šiuolaikinėje muzikoje* by the Lithuanian Academy of Music and Theatre (Vilnius: Lietuvos muzikos ir teatro akademija, 2013, ISBN 978-609-8071-10-8)

Supported by the Lithuanian Council for Culture, the Research Council of Lithuania, and the Lithuanian Academy of Music and Theatre

Mathematical editing by Dr. GIEDRIUS ALKAUSKAS
Translated by LAIMA VINCĖ SRUOGINIS
Language editing by KENNETH BASFORD

ISSN 1618-842X
ISBN 978-3-631-71381-5 (Print)
E-ISBN 978-3-631-71382-2 (E-PDF)
E-ISBN 978-3-631-71383-9 (EPUB)
E-ISBN 978-3-631-71384-6 (MOBI)
DOI 10.3726/b10528

PL Academic Research is an Imprint of Peter Lang GmbH.

Peter Lang – Frankfurt am Main · Bern · Bruxelles · New York · Oxford · Warszawa · Wien

This publication has been peer reviewed.

www.peterlang.com

Contents

Part 2
The Renewal of Mathematical Techniques in Musical Compositions of the 20th and 21st Centuries

List of Analyzed Music Scores

Author's Preface

The concept of *musica mathematica*[1] seeks to accurately examine the intersection of two seemingly radically different subject areas. In the sphere of musical research, this concept is in essence related to one of the trends of music – attributing the theory of music to science – to the sphere of *mathematica*. Over the course of different epochs, musical theory has been classified as Latin *musica theorica/ theoretica/contemplativa/speculativa/arithmetica* etc. Theory was the domain of the scientist who came armed with academic knowledge and who studied the esoteric secrets of music.

From the perspective of a European perception, the definition of the science of music was a result of the Pythagorean concept, which is based on the universal harmony of numerical proportions. The Pythagoreans were the first in European culture to raise the issue of uniting music and mathematics, sound and number. They perceived music as an abstract sphere based on mathematical means.

In the Middle Ages, Boethius (c. 480–524/5) considered this problem in his treatise *De institutione musica* (6th century). He placed music in the quadrivium of the mathematical sciences, thus continuing the position of the thinkers of Antiquity. The influence of this position was also reflected in later treatises, for example in the study of the 13th century *Musica speculativa secundum Boetium* (1323) of the French mathematician, astronomer, philosopher and music theorist Johannes de Muris (1290–1351/5), who taught at the Sorbonne. It is also reflected in works of the 17th century, titled *musica mathematica*. They are *Musica mathematica* by the German physician, astronomer and mathematician Heinrich Brucaeus (1530–1593)[2] or the study *Dissertatio mathematica de musica* (1672) by Johann Christoph Wegelin (1650–1726) and the mathematician and astronomer Julius Reichelt (1637–1717), who taught at the University of Strasbourg. However, the source of *musica mathematica* that was of special significance in that century was the treatise *Musica mathematica* (1614) by Abraham Bartolus. This treatise analyzed – alongside different manners of tuning – the tuning of sounds proposed by Andreas Reinhard – dividing the Phrygian scale into 48 equal parts as an analogue of cosmic proportions. His insights influenced later representatives of musical theory, for example Johann Mattheson's (1681–1764) considerations about the relationships between music and mathematics.

---

1 Latin – mathematical music, musical mathematics.

2 Joachim Burmeister, who prepared and published the treatise by Brucaeus in 1609, named it more comprehensively – *Musica theorica*.

The idea of the art of sound being based on mathematics referred to by Gioseffo Zarlino (1517–1590) as *numerus sonorus* or *numerus in sono* (Latin – sounding numbers) has been developed for centuries. It has been called the longest lasting interdisciplinary dialogue with the original imprints of mathematical composition of music in every epoch. This dialogue is especially concentrated in the practice of composing contemporary music. Here the synthetic phenomenon of music and mathematics plays a very important role, acquiring an interdisciplinary trend of scientific research in which the creation of music is characterized as a qualitatively new stage of contact between these spheres. This has markedly expanded the boundaries of mathematical expression in music and has integrated not only the diversity of the phenomena of musical numerology (for example, ideas about the nature of cosmological musical intervals or the constructivism of numerical proportions and progressions, the application of combinatorics of music or the creation of musical messages encoded in numbers, etc.) but has also engendered innovative ideas in the sphere of mathematics and the preconditions of musical and numerical structures when mathematical laws of the group theory become "common denominators" and the logic of Markov chains is interpreted as the law of melodic processes. In addition, practices of tuning are based on specific mathematical theories, algorithms generated by computer are implicated in the space of the sounds, the process of composing is based on very formalized idioms of a mathematical nature. Personality and creative exceptionality is also very pronounced in this current colorful creative panorama – modern composers choose, manipulate, and use mathematical technologies or traditionally determined semantics of musical numerology in a highly individual manner.

I have dedicated more than fifteen years of my life to the study of *musica mathematica*. I was motivated by my aspiration to become acquainted with, and to comprehend, the interaction between art and science (or between two sciences). In this monograph, I seek to systematize the research methods of this phenomenon while considering the retrospective of historically formed numerological musical traditions. I will present a range of diverse mathematical laws in the practice of music composition in the 20th and 21st centuries.

In the three parts of this monograph, I display versatile cases of the intersection of music and mathematics, moving gradually from philosophical and aesthetic considerations about *mathesis* as the idea of universal beauty to practical studies of the implementation of this idea. From there, I move to a discussion of the interaction between music and other kinds of art (architecture, painting, poetry and literature) by making use of the common denominator of mathematics, and practical research, as well as the expression of a modern attitude towards a thousand-year long dialogue. I will also take a close look at the innovations of the turn of the 20th and 21st centuries.

Another peculiarity of the structure of this monograph is that it shows commonalities between *musica* and *mathematica.* For example, a dichotomous attitude to this phenomenon determined the generalization of the traditions of mathematical techniques of earlier periods, their renewal in the music of the 20th–21st centuries, and modern innovations according to purely formal-constructive and notional-symbolic ways of rendering musical compositions mathematically. This monograph considers theoretical and compositional cases of contact between the composition of contemporary music and mathematics. My work examines the manner in which this phenomenon effects the process of composing; what musical parameters are most optimal for mathematical implications; thereby what forms of specific mathematical procedures are introduced in contemporary compositional texts, and how all that influences the resulting musical composition. I have used systematic, structural and complex methods of analysis in: studying scientific material related to the compositional expression of numerological manipulations and identifications of mathematical aspects or strategies for their implication; grouping and systemizing documents of different epochs of philosophical and compositional intersections of music and mathematics; defining the theoretical aspects of music in terms of mathematical equivalents; analyzing selected musical compositions, at the same time showing that rendering compositional texts mathematically is not a mechanical and formal process.

My research that has resulted in this monograph is only a small step in a thousand-year-long dialogue between music and mathematics. However, I am grateful to Professor Gražina Daunoravičienė for guiding me in this direction and for her invaluable insights and advice, which I needed when moving deep into the space of *musica mathematica.* I am grateful to my husband, Girėnas, for his support and patience. I would like to thank everyone who has helped me write this monograph.

Foreword: *Mathesis* as a Philosophy of the Beauty of Music

Most likely, one of the answers to the question where the vitality of the idea of the interaction between music and mathematics – which can be traced for more than two thousand years and which has been developed in the modern world – lies in the perception that mathematics is the principal cause and source of an all-embracing beauty. The idea of a mathematically substantiated world was developed as far back as Antiquity, and through its expression has attracted thinkers of this epoch. The idea was in line with the worldview of the later epochs as well: Galileo Galilei (1564–1642) called mathematics the alphabet by means of which God describes the world.[3] The Hungarian mathematician Paul Erdős (1913–1996) spoke about the imaginary divine book that contained the most beautiful mathematical proofs. The following utterance is attributed to Erdős:

> [When asked why numbers are beautiful?] It's like asking why Ludwig van Beethoven's Ninth Symphony is beautiful. If you don't see why, someone can't tell you. I *know* numbers are beautiful. If they aren't beautiful, nothing is.[4]

The Russian philosopher Alexei Losev (Алексей Фёдорович Лосев, 1893–1988) related the definition of beauty to numbers as well. He stated that beauty was something "impersonal" that was neither spirit nor personality, but an impersonal, non-qualitative structure. He cited, as examples, numbers (or an atom, the initial

---

3 The quotation in Italian "La mathematica è l'alfabeto in cui Dio à scritto l'Universo" is said to have been recorded by Galileo Galilei in his treatise *Il Saggiatore* (Rome) in 1623.

4 Cited in Paul Hoffman: *The Man who Loved only Numbers. The Story of Paul Erdős and the Search for Mathematical Truth*, London, 1998, p. 44. Bruce Schechter, the author of Erdős' biography, states that the eccentric Hungarian mathematician was an atheist but speaking about an especially successful mathematical proof he used to say that it was recorded in The Book:

> Perhaps Erdős' most interesting coinage is the term "Supreme Fascist", or SF, which is what he called the God in whom he professed no belief. […] The SF is the author of The Book of all the best mathematical proofs, and it is one measure of His cruelty that He keeps its content hidden. We are therefore obligated to use all of our intelligence and intuition to reproduce the contents of the SF's hidden Book for ourselves. (Bruce Schechter: *My Brain is Open. The Mathematical Journeys of Paul Erdös*, New York: Simon & Schuster, 1998, p. 70)

structure, “the primary kernel”, “the clasp of the entire construction”; see Losev 1963: 506–7 & 1999).

The aesthetic concept of mathematics as an expression of universal beauty determined not only philosophical considerations, but also practical investigations. These concentrated on searching for harmony hidden within the depths of this science, and on the attempt to define what provided the possibility of referring to mathematical processes as aesthetic.

What makes one mathematical proof considered as more attractive than another, regardless that both are correct? Elegance and refinement are inseparable from mathematical calculations and make the latter something more than mechanical manipulations with symbols. For example, one way of proving a theorem could be characterized as a boring theoretical operation, whereas another could be defined as a manifestation of elegance, in which every step of the calculation is a process, similar to that of a creation of a work of art.[5] Herbert E. Huntley, who investigated the beauty of the Golden Ratio[6] stated that the emotions of a scientist who has proving a theorem are identical to those experienced while admiring a masterpiece of art, because both are a creative work (Huntley 1970: 76). According to the German musicologist Heinrich Husmann, who brought attention to a particular interest in the number that prevailed in Antiquity and admiration for the infinite possibilities of its manipulations, the ancient Greeks found arithmetic proportions less interesting than harmonious relations between numbers.[7] This explains in part why mathematicians are intrigued by irrational numbers, for example the Archimedes constant π, which is thought to be beyond the powers of human perception, or the story about the swift-footed Achilles who never catches up with the turtle.[8]

---

5 For example, at the age of three, the mathematician Carl F. Gauss (1777–1855) solved the puzzle of what all the numbers added together from 1 to 100 would equal without doing any summation operations, but by having noticed that pairs of numbers 1 and 100, 2 and 99, 3 and 98, 4 and 97, etc. were equal to the same sum 101, which had to be multiplied 50 times (Rothstein 1995: 143).

6 Also Golden Section or Golden Mean, Latin *sectio divina*, *sectio aurea* – divine section, golden section.

7 Husmann reasoned this by presenting an example of numbers 6 and 8: their arithmetic mean 7 is finite, so the arithmetic proportion is directly mechanical and not inspiring. The harmonic mean of these numbers $6\ {}^{6}/_{7}$, on the contrary, gives an impetus for a further development of numerical operations (more see Husmann 1961).

8 This is the so-called Zeno’s paradox of continuum, attributed to pre-Socratic Greek philosopher Zeno of Elea (Ζήνων ὁ Ἐλεάτης, c. 490–c. 430 B.C.). If the tortoise is the first to move from point A to point B, Achilles, who starts later moving from the

The attitude towards mathematics as the art of the beauty of numbers, which has prevailed since Antiquity, has had an effect on the environment as well: operations with numbers, regularities of symmetry and proportions, have become beauty formulas in different spheres of art. For example, *beautiful* in music is often perceived as that which is hidden in relation to mathematics (Powel 1979: 267). The logic of the structure of music is often identified with the logic of a mathematical equation. A musical composition is defined as "a dramatically, passionately told mathematical story" (Hammel 2015). St. Thomas Aquinas' (1225–1274) phrase deserves mention within this context too – "music, which studies the ratios of audible sounds."[9] The later statement by Gioseffo Zarlino (1517–1590) that "music derives its principles from natural science and from [the science] of numbers" is made in the first book of his treatise *Le istitutioni harmoniche* (1558).[10]

The fundamental theory of Alexei Losev stands out for its philosophical considerations about the relationship between music and mathematics. The art of sounds is clearly separated from music as a physical or psychic phenomenon in that theory. According to Losev, "only ideal relations of numbers can equal to the art of sounds", because the nature of music and mathematics is *sui generis*. Alogical thinking, however, is characteristic of art, and it surpasses the logical thinking of mathematics. This is because mathematics is only a construction of numbers and speaks about the number in a logical way, whereas music expresses the number *artfully*; it does an expressive and symbolic construction and reveals the artistic side of the number (Losev 1999).

---

same point, by following the tortoise cannot catch up with it and every time lags behind it at a certain distance – when Achilles reaches point B, the tortoise has already travelled to point c. When Achilles reaches point C, the tortoise has already moved to point D, etc. The analogous example can be that of infinite process of summation when trying to calculate number 1: it is obvious that 1 = 1/2 + 1/2 or 1 = 1/2 + 1/4 + 1/4 or 1 = 1/2 + 1/4 + 1/8 + 1/8. However, the process is infinite when 1 = 1/2 + 1/4 + 1/8 + 1/16 + 1/32…

9 Original quote in Latin "Musica, quae considerat proportiones sonorum audibilium" was written in St. Thomas Aquinas' *Commentary on Aristotle's Metaphysics* (Bk 3 Lsn 7 Sct 412, transl. John P. Rowan, Chicago, 1961, p. 201).

10 Cited from Zarlino's treatise *Le istitutioni harmoniche*, Chapter 20 "On Why Music is Subject to Arithmetic and is the Intermediary between Mathematics and Nature" (Zarlino 1562: 30), also see the translation by Lucille Corwin (*Le Istitutioni Harmoniche" of Gioseffo Zarlino, Part 1: A Translation with Introduction*, ProQuest, UMI, 2008, p. 324). This chapter discusses the relations between music and arithmetic on the basis of the famous Persian polymath Avicenna (also known as Ibn Sīnā, c. 980–1037).

A non-accidental union of music and mathematics is also obvious in their fundamental commonalities. Music and mathematics is articulated in the language of symbols, graphical signs. However, according to Marcus du Sautoy:

> Music is much more than just the minims and crochets which dance across the musical stave. Similarly, mathematical symbols come alive only when the mathematics is played with in the mind. As Pythagoras discovered, it is not just in the aesthetic realm that mathematics and music overlap. The very physics of music has at its root the basics of mathematics. (Sautoy 2003: 78)

The mathematician Euclid (Εὐκλείδης, c. 365–275 B.C.) who is called the father of geometry, and almost two thousand years later, Galileo Galilei, were both concerned with the same question: why combinations of some sounds are more beautiful than others (Rothstein 1995: xvi). Gottfried Leibniz (1646–1716) was one of the first to speak about the links between prime numbers[11] and music. While trying to establish the law of the sequence of prime numbers, he looked for causes and answers in the process of a musical composition (Klotz 2006: 105). The mathematician Leonhard Euler (1707–1783), who believed that prime numbers lay behind certain beautifully sounding combinations of tones, investigated relations between the consonants of music and integers (Rothstein 1995: xvi). Marcus du Sautoy refers to the 19th century mathematician Georg F. B. Riemann (1826–1866), stating that nature may have hidden in the primes the music of some mathematical orchestra (Sautoy 2003: 93, 97).[12]

The opposite opinion is also to be mentioned: for example, the philosopher, mathematician and historian Bertrand Russell (1872–1970) characterized mathematics as having "cold and austere beauty" and separated it radically from the art of music due to smartness characteristic of the latter. The British thinker saw parallels between mathematics and poetry:

> Mathematics, rightly viewed, possesses not only truth, but supreme beauty – a beauty cold and austere, like that of sculpture, without appeal to any part of our weaker nature, without the gorgeous trappings of painting or music, yet sublimely pure, and capable of a stern perfection such as only the greatest art can show. The true spirit of delight, the

11 The characteristic feature of the prime numbers is that they have no divisors, with the exception of 1, that is, they are divisible only by 1 or by themselves.

12 According to Sautoy, music was the first impetus that aroused interest of mathematicians in the possibilities of the prime numbers and their infinite sequence. This is related to the discoveries by ancient Greeks of the infinite division of the string at the ratios 1/2, 1/3, 2/3, 3/4, etc. seeking to obtain musical consonance. This associated with the infinity of the series of the prime numbers (Sautoy 2003: 77).

> exaltation, the sense of being more than Man, which is the touchstone of the highest excellence, is to be found in mathematics as surely as poetry.[13]

The point of intersection of commonalities between music and mathematics is analogous or isomorphic actions/processes. For example, the table of transpositions of musical tone-series (series of the original O, its transpositions – inversion I, retrograde R and retrograde inversion RI) is isomorphic to Klein four-group.[14] Variants of transformations of geometric objects and musical transpositions are analogous:

- the most elementary way of geometric transformation is geometric translation. It corresponds to a simple repetition of a musical element (O);
- geometric transposition corresponds to a musical sequence (O);
- geometric reflection corresponds to musical retrograde (R);
- geometrical inversion and glide reflection correspond to musical inversion (I);
- geometric rotation corresponds to the inversion of musical retrograde (RI).

---

13 The statement by Rusell first appeared in the Chapter 4 "The Study of Mathematics" of his *Mysticism and Logic, and Other Essays*, London, 1919, p. 60. Later the quotation was widely cited by various authors.

14 Klein four-group is named after the mathematician, initiator of the theory of geometry groups Felix Klein (1849–1925), is a subgroup of the symmetric group S_4. It consists of four elements with which permutations are performed: (), (12)(34), (13)(24) and (14)(23). Four combinations of elements can be demonstrated by means of transformations of the rectangle, which is not a square (see example below): the first one is the original position of the rectangle (*e*), the second one is a reflection of the rectangle through the horizontal axis (*f*), the third permutation is a reflection through the vertical axis (*g*), and the fourth permutation is a 180° turn of the rectangle through the center (*h*). These operations can be reduced to a table, which shows in what position the rectangle finds itself when performing several permutations:

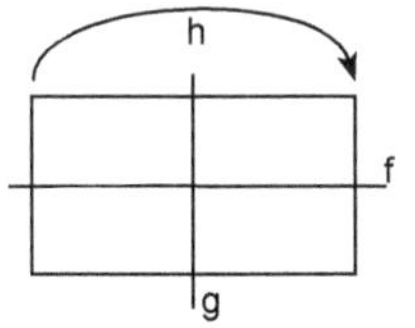

x	e	f	g	h
e	e	f	g	h
f	f	e	h	g
g	g	h	e	f
h	h	g	f	e

It has been noticed that transposition operations characteristic of Klein four-group correspond to transpositions of musical tone-series:

	O	I	R	RI
O	O	I	R	RI
I	I	O	RI	R
R	R	RI	O	I
RI	RI	R	I	O

The mathematical terminology used in the 20th century to define the processes of music composing testifies to the affinity between music and mathematics. For example, the terms *system* and *mapping* were used by the composers Milton Babbitt, Otto Laske, and Barry Truax. The pitch class theory uses the system of modules to mark the pitch. For example, the class of sounds within the octave is marked modulo 12 (mod12), each sound is respectively written down in terms of the whole number:

$$C = 0 \pmod{12},\ D = 2 \pmod{12},\ E = 4 \pmod{12}, \text{ etc.}$$

The idea of an interaction between music and mathematics that is consistently developed, and which has left its mark in every epoch, is of great importance in the contemporary world as well. In general, in the realm of contemporary music, the constructive manipulation of the interaction between music and mathematics has become especially active. Fred Lerdahl and Ray Jackendoff based their music theory on mathematics as a constructive basis. These insights were adapted from Heinrich Schenker's *Ursatz* theory (score reduction and the process of prolongation). Mathematical formalism was typical of David Lewin's theory presented in the early 1980s about the conceptual space of music, its points/elements and distances/intervals between them. In Lewin's theory, musical space occupied all three dimensions – pitch, rhythm, and timbre. Each of them could be placed in structural models. Intervals between groups could be measured with a mathematical group system, called *Generalized Interval System* (GIS) (more see Lewin 1991).

Contemporary composers confirm the close tie between music and mathematics. Ernst Křenek (1900–1991) compared musical thinking with the independence of axioms. Pierre Boulez (1925–2016) compared it with mathematical thinking and with thinking about physics. George Crumb (born 1929) described music as a "system of proportions that serves a spiritual impulse" (Takenouchi 1987: 1). Witold Lutosławski (1913–1994) saw music and mathematics as two complicated sciences with something in common. A variety of mathematical operations became the inspiration for the compositions of Iannis Xenakis (1922–2001).

Having spread as the accumulation of experience of a two thousand-year old tradition, the idea of *musica mathematica* determined very diverse mathematical aspects of the music composition of the 20th–21st centuries. Going deep into the sources and development of this interaction is of great importance in analyzing them. Therefore, in the first part I seek to reveal confrontation encoded in the nature of music and mathematics, the issue of constant comparison and interlacement, and the multi-aspect nature of this interaction during different periods in a retrospective analytical way. The second and third parts present a wide spectrum of the relationship between music and mathematics while analyzing musical compositions of the 20th and 21st centuries.

Part 1
A Retrospective of the Traditions of *Musica Mathematica*

1. The Constructive Relationships between Music and Mathematics: The Pythagorean Conception of Universal Music and its Spread in the Worldview of Later Periods

The popular concept that "there is geometry in the humming of strings, there is music in the spacing of the spheres" is attributed to Pythagoras (Πυθαγόρας, c. 570–497/6 B.C.).[15] The name of Aristotle (Ἀριστοτέλης, 384–322 B.C.) is related to a scholastic definition of beauty: the criteria of beauty, "orderly arrangement, proportion, and definiteness", are "especially manifested by the mathematical sciences."[16] The last of the three most famous authors of Greek tragedy, Euripides (Ευριπίδης, c. 480–406 B.C.), said that "mighty is geometry; joined with art, resistless."[17]

Statements made by the polymaths of Antiquity testify to the fact that the fundamental principles of the philosophic idea about the interaction between music and mathematics were formed as far back as ancient Greece. This idea naturally resulted from the perception of the all-surrounding environment that was made mathematical, and its materialistic substantiation, which dominated in the antique world outlook and which was given special significance in the works of the philosophers of the 6th–4th centuries B.C., underlying "perfect knowledge of canonical rules and the ability to base oneself on them in one's creative work" (Andrijauskas 1996: 294). The physical sensual experiences of an antique man, a material comparison or contrasting of things, was adapted to getting to know the environment – the cosmos using human's senses united into *ratio*.[18] This concept gave rise to the theory about the number as the main constructive element of universality; at the beginning of the year of the Lord, the theory was continued by

15 Pythagoras' quotation was cited in Louise B. Young's *The Mystery of Matter* (Oxford University Press, 1965, p. 113).

16 The definition was presented by Aristotle in his *Metaphysics*, Book 13, Part 3, 1078a–1078b (transl. Hugh Tredennick); also quoted in Albert L. Blackwell's *The Sacred in Music* (Westminster John Knox Press, 2000, p. 162).

17 Euripides' quotation was cited in *Mathematically Speaking: A Dictionary of Quotations* (eds. C.C. Gaither & Alma E. Cavazos-Gaither, CRC Press, 1998, p. 84).

18 Greek *νους*, Latin *ratio* – reason.

St. Augustine (354–430). Admiring the eternity and universality of the number, he stated the following:

> But seven and three are ten not only at the moment, but always; [...] this incorruptible numerical truth is common to me and to any reasoning being.[19]

1.1. Music and the Theory of the Quadrivium

The idea that a group of mathematical sciences united music, astronomy, geometry and arithmetic belongs to the Pythagoreans. These four sciences were called quadrivium by Boethius (c. 480–524/5)[20] in the 6th century and were included in the system of seven disciplines – *septem artes liberales* – taught at universities in the Middle Ages. In the 9th century, Aurelian of Réôme was the first author to have transferred Boethius' mathematical theory of music into a practical space in his treatise *Musica disciplina*[21] – into a Gregorian monody: in Chapter 2 of the treatise the author writes about the mathematical proportions of musical intervals and chooses examples of songs and chants to illustrate the intervals.

Figure 1. A graphical expression of septem artes liberales

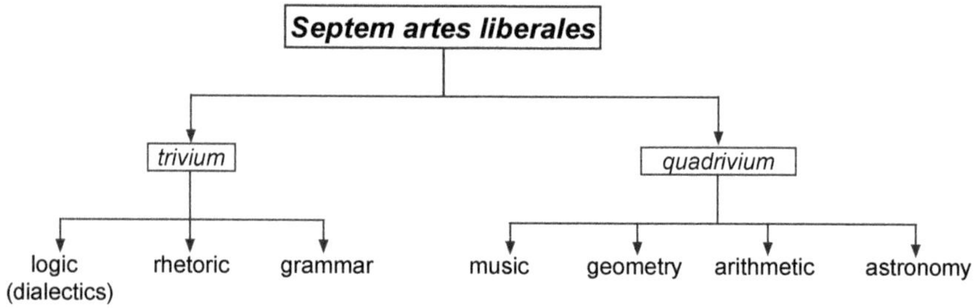

19 St. Augustine's statement was published in his *De libero arbitrio* (On Free Choice of the Will, in 3 volumes, written in 387–395, second book 2.8.21.83). The quotation was cited in English translation by Peter King (Cambridge University Press, 2010).

20 Boethius also referred to as Anicius Manlius Severinus Boëthius.

21 Aurelian of Réôme, also referred to as Aurelianus Reomensis (fl. c. 840–850). His treatise *Musica disciplina* of the ninth century is considered the earliest surviving medieval treatise about music; the exact date of writing it is unknown.

Figure 2. Medieval drawings representing the septem artes liberales*: 1–3* trivium *of grammar, logic (dialectics) and rhetoric, 4–7* quadrivium *of geometry, arithmetic, music and astronomy*[22]

22 The medieval drawings are taken from Thomas von Zerklaere: *Der Wälsche Gast*, Ostfranken (?), 1340, fol. 65v. Manuscript at Forschungsbibliothek Gotha, Cod. Memb. I 120, http://digi.ub.uni-heidelberg.de/diglit/fbg_membI120 [accessed 12 September 2016]. Published with kind permission of the Forschungsbibliothek Gotha at the University of Erfurt.

Attributing music to the sphere of mathematics was characteristic of the Baroque epoch as well. The type of music found at that time, *musica scientia* (also *musica contemplativa/speculativa/theorica/theoretica*), was called by Gottfried Leibniz, Johannes Lippius, Jakob Adlung or Andreas Werckmeister "sounding mathematics", "mathematical knowledge", "the daughter of mathematics" or "the science of mathematics creating harmonious singing" (Lobanova 1994: 128). For example, in the first dictionary of music written in the German language, *Musicalisches Lexicon Oder Musicalische Bibliothec* (Lexicon of Music or Music Library, 1732), whose author is Johann Gottfried Walther (1684–1748), among different types of music, the type *musica arithmetica* is indicated, that is, arranging the sounds in proportions and numbers (Walther 1732: 431). Johannes Kepler (1571–1630) called the theory of music perfect science (German *vollkommene Wissenschaft*), whose eternal truth is mathematics.[23] In his treatise *Plus ultra* (1754–1756), Johann Mattheson (1681–1764) identified the process of composing music with an architectural construction and advance calculations peculiar to it. In 1743 in the monthly *Neu eröffnete musikalische Bibliothek* (The New Music Library) Lorenz Mizler (1711–1778) stated that mathematics is the heart and soul of music, and:

> Without question the bar, the rhythm, the proportion of the parts of a musical work and so on must all be measured. [...] Notes and other signs are only tools in music, the heart and soul is the good proportion of melody and harmony. It is ridiculous to say that mathematics is not the heart and soul of music.[24]

In the preface to the cycle of six sonatas for the clavier *Musicalische Vorstellung einiger biblischer Historien* (Musical Representation of Several Biblical Stories, 1700) the composer Johann Kuhnau (1660–1722) wrote the following:

> I should investigate further the fundamentals of music and demonstrate the splendid and wonderful usefulness of *Mathesis* and in particular the *ars combinatoria*, which serves the inspiration so magnificently [...].[25]

1.2. Review of the Harmony of Spheres

In Antiquity the environment (cosmos) that surrounded man was treated as a creation of God. The cosmos was an example of perfect order or harmony[26] due

23 Quotation in German "in der Mathematik ihre principia aeternae veritatis hat" was cited in Heher 1992: 30.

24 Mizler's statement was published in *Neu eröffnete musikalische Bibliothek*, Bd. 2, Leipzig, 1743, p. 54; English translation quoted from Tatlow & Griffiths 2016.

25 English translation by Michael Talbot, cited in Kuhnau 1700/1973: xii.

26 Greek *harmonía/αρμονια* – connection, union, sounding, concord of sounds.

to the characteristics of regularities, symmetry and proportions, which were expressed in certain numerical relations.[27] This explanation of universal harmony in terms of numbers, and making mathematical principles absolute, determined the originality of the aesthetics of the Pythagoreans. Laws of the cosmos order were related to the primary conditions that were necessary for man's existence – the four elements: earth, air, water and fire,[28] to their interaction and contrasting,[29] as well as to *arithmos*[30] expressing their relations. This is testified to by the work of antique literature written in c. 361 B.C. – Plato's dialogue *Timaeus*.[31] This literary work talks about the Cosmos as about a body: four elements are necessary for matter to exist, harmony between them is constructed in accordance with the relations between three main proportions – those of arithmetic, geometry and harmony,[32] which were determined by a systematic change of three pairs having

---

27 This is encoded in the very meaning of the word "cosmos": Greek *cosmos/κόσμος* – order, orderly arrangement, a harmonious system.

28 Greek: *στοιχεῖον* (four elements), *γαῖα*, *γη* (earth), *ἀήρ* (air), *ὕδωρ* (water) and *πυρ* (fire). Cf. the fifth element – a tree – is also mentioned in the oriental culture.

29 Four elements formed the basis of the theory of the thinkers of the antique epoch about the creation of the World. For example, Thales of Miletus (Θαλης ο Μιλήσιος, c. 624–c. 546 B.C.) looked for the rudiments of philosophy in water, which he considered to be the symbol of wisdom, deity and life; in his theory of the appearance of life Anaximander (Ἀναξίμανδρος, Anaximandros, c. 610–c. 546 B.C.) stated that the rudiments of life were in the union of two elements – *land* and *water* (*On Nature*, c. 547–6 B.C.); in the opinion of Heraclitus of Ephesus (Ηράκλειτος ο Ἐφέσιος, c. 535–c. 475 B.C.), the world resembled a flame, the latter was not only an all-destroying, but also an all-creating power.

30 Greek *αριθμος* – number.

31 Plato's (Πλάτων from Athens, 428–347 B.C.) dialogue *Timaeus* (*Τιμαιος*) was written following the author's third journey to Sicily c. 361 B.C.

32 1) in ARITHMETIC proportion, the second dimension is larger than the first one as much as the third dimension is larger than the second:
1 : 1 ½ : 2 or an equivalent in the sequence of the cosmic seven-term 1 : 2 : 3.
This is a proportion of numbers that was considered to be the simplest one in Antiquity whose progression can be expressed in terms of the relationship between the third and the second magnitudes. For example, 0–1–2–3–4–5–6–7–8–9… or 0–2–4–6–8–10–12–14 …
2) in GEOMETRIC proportion, the ratio between the second and the first magnitudes coincides with that of the third and the second magnitudes:
1 : 2 : 4, because 1 : 2 = 2 : 4.
The progression of numbers is formed accordingly: 1–2–4–8–16–32–64–128 …
3) HARMONIC proportion expresses the relationship between three numbers when the third number is larger than the second one by such part of its size as the second

Figure 3. The relations between the four elements in Antiquity

FIRE	AIR	WATER	EARTH
subtle	subtle →	dense	dense
sharp →	blunt	blunt	blunt
mobile	mobile	mobile →	immobile

material properties: "subtle–dense", "sharp–blunt" and "mobile–immobile." This text by Plato can be transferred into a scheme showing a consistent transition from one substance into another with only one property changing (see Figure 3).

Plato's dialogue is a work about a mathematically calculated idea of beauty and the importance of numerical relations to the creative process. In presenting the theory of the creation of Cosmos, the author states that Demiurge, a divine Craftsman,[33] when creating the world according to his own image ("He desired that all things should be as like himself as they could be"; *Timaeus*, 30a) performed certain mathematical operations. This is shown by the extract cited below about the "making" of the cosmic soul from a mixture divided according to the algorithms of the so-called cosmic septenarius 1–2–3–4–8–9–27 and three proportions:[34]

> And he proceeded to divide after this manner: – First of all, he took away one part of the whole [1], and then he separated a second part which was double the first [2], and then he took away a third part which was half as much again as the second and three times as much as the first [3], and then he took a fourth part which was twice as much as the second [4], and a fifth part which was three times the third [9], and a sixth part which was eight times the first [8], and a seventh part which was twenty-seven times the first [27]. After this he filled up the double intervals [i.e. between 1, 2, 4, 8] and the triple [i.e. between 1, 3, 9, 27] cutting off yet other portions from the mixture and placing them in the intervals, so that in each interval there were two kinds of means, the one exceeding and exceeded by equal parts of its extremes [as for example 1, 4/3, 2, in which the mean 4/3 is one-third of 1 more than 1, and one-third of 2 less than 2], the other being that kind

---

number is larger than the first one by the same part of the size of the first; that is, these three numbers are a, $\frac{2ab}{a+b}$, b:
1 : 4/3 : 2 or 3 : 4 : 6.
The harmonic proportion intrigued the thinkers of Antiquity most, because it was special, as its algorithm allowed fractional relationships of numbers to be compared.

33 Greek *dēmiourgos*/*δημιουργός* – craftsman, creator.

34 Quote from Plato's *Timaeus*, 35b–36b, translated by Benjamin Jowett (published in: *The Dialogues of Plato*, Vol. 3: *The Republic, Timaeus, Critias*, 3rd ed., Oxford University Press, London: Humphrey Milford, 1892, p. 36).

of mean which exceeds and is exceeded by an equal number. Where there were intervals of 3/2 and of 4/3 and of 9/8, made by the connecting terms in the former intervals, he filled up all the intervals of 4/3 with the interval of 9/8, leaving a fraction over; and the interval which this fraction expressed was in the ratio of 256 to 243.

Plato's dialogue, however, is significant not only as a confirmation of the mathematical nature of the world. This written source of the 4th century B.C. established the idea of the structure of the Cosmos based on harmoniously sounding, rather than simple and "dry" calculations. Having compared the creative process described by Plato and the Pythagorean system of numerical relations of musical intervals, it becomes clear that music participated inseparably in the creation of the world. The proportions of numbers in Plato's dialogue coincide with the theory about recording musical intervals in terms of numerical formulas created by the Pythagoreans.[35] For example, the proportions of "decomposing the mixture" used in Plato's description of Demiurge's work are identical to the numerical relations of musical intervals – the perfect octave 2 : 1, perfect fifth 3 : 2, perfect fourth 4 : 3 and whole tone 9 : 8 (3/2 : 4/3) whose common basis is the numerical sequence 1–2–3–4 of Pythagoras' perfect numbers, the so-called tetractys. The mystification of tetractys reveals itself in the words of the Neoplatonist philoso-

35 The history of a numerical substantiation of musical intervals is rephrased by a legend about Pythagoras: having heard different consonances produced by hammers, which were coming from the smithy, Pythagoras established that respective relationships of hammers of different weight determined a certain sound of musical intervals and that relationships between natural sequences of the first numbers determined the most harmonious consonances of the sounds.
The musicologist Heinrich Husmann recreated the process of how the ancient Greeks created the numerical equivalents of musical intervals on the basis of a one-string instrument called monochord (Greek *μονόχορδος* – one string). Taking into account the spot where a finger was put on the string, a concrete interval was obtained: having divided the string in half 2 : 1, the interval of the octave sounded; having put a finger in the place at two thirds of the string 3 : 2, the interval of the fifth sounded, and so forth. Relationships between other musical intervals forming an equal-tempered scale were derived on the analogous principle (Husmann 1961: 9–19).
In c. 1700, ancient scale temperament was started to be considered imperfect, and the so-called "perfect tuning" was established (Apel 1970: 835–6). But before forming the well-tempered scale of 12 semitones, other attempts in the theory of music are to be mentioned, for example in the middle of the 17th century Nicola Vicentino proposed the just intonation system and division of the octave into 31 parts; Gioseffo Zarlino and Francisco de Salinas proposed to divide it into 19 equal parts, etc.

pher Iamblichus (’Ιάμβλιχος, c. 245–c. 325, also known as Iamblichus Chalcidensis, or Iamblichus of Apamea):

> What is the oracle at Delphi? The tetractys. What is harmony? That in which the Syrens subsist. (Iamblichus 1818: 43)

Iamblichus testifies to the fact that proportions made from its numbers 1, 2, 3, and 4 in Antiquity were a sign of perfection and cosmic stability, since these three musical intervals formed with the help of these numbers were interpreted as euphonious accords – consonances (octave 2 : 1, fifth 3 : 2 and fourth 4 : 3). This encouraged numerical proportions to be called consonance.[36]

The numbers of the tetractys in the antique worldview were used to count changes within the four classic elements: fire, air, water, and earth. Therefore,

36 The secret of tetractys is showed in the pyramid of dots whose graphical expression is an isosceles triangle:

● 1
● ● 2
● ● ● 3
● ● ● ● 4

A graphical expression of tetractys also represented the analogy of colors and the sounds of music: three points at the top of the pyramid is a threefold bright white color as the symbol of divinity, the other seven points are seven colors of the spectrum that correspond to seven sounds of music: *c* – red, *d* – orange, *e* – yellow, *f* – green, *g* – blue, *a* – indigo, and *b* – violet or purple (Hall 1928: 84).
In the opinion of Carl Dahlhaus, the traditional definition of the perfection of tetractys was somewhat “mislead” by a discrepancy – this is the perfect eleventh (spanning an octave and a fourth), being perfect it was expressed by one non-tetractys number 8, because the perfect eleventh is 3 : 8 (Dahlhaus 1985: 18). Therefore, the thinkers of Antiquity, seeking to legalize consonance of perfect eleventh (or compound fourth), alongside the traditional four of the first numbers 1–2–3–4 mentioned another influential fourth of numbers 6–8–9–12 which contained terms necessary to the proportions of perfect twelfth (or compound fifth) as well. This sequence of numbers was called musical tetractys, which also expressed the possibilities of three proportions – arithmetic, geometric and harmonic.
In the 16th century boundaries of tetractys were expanded up to the hexachord – senario (sequence of numbers to 6). Having done that Zarlino motivated that number 6 was considered to be the first perfect number (equal to the sum of its divisors) and pointed to six stages of man’s age, six planets as senario analogues. In this way it was sought to legalize other intervals as perfect (consonant) – the minor and major thirds (6 : 5 and 5 : 4), as well as the major sixth (5 : 3) that formed within the limits of senario (Dahlhaus 1985: 23).

the Pythagoreans' digital definition of musical intervals became the justification that unified harmony in the world. The Cosmos of the antique world was "melded" or "connected" through various elements in an orderly fashion, which created sound in perfect musical intervals. These matches influenced the material proportions of the four elements, because, according to the Pythagoreans, the earth was made up of four of its own parts; water was made up of three parts earth and one part fire. Air was made up of three parts fire and one part earth, while fire was made up of four of its own parts (see Figure 4).

Yuri Cholopov noted that in Antiquity numerical proportions reflected the structure of the heavenly spheres according to the plan of cosmic music heptachord (system of seven sounds), and that the Universe was perceived as the prototype and model of the harmony of sound (Cholopov 1988: 190). Therefore, the relationship of musical tones expressing the proportions of numbers were compared with characteristics of the Cosmos, so that it would be possible to "measure" the world of the micro (the human body) and macro (the Universe), as interpreted as a taut string,[37] divided into perfect intervals. This concept was named the Harmony of the Spheres.[38] According to this principle, the positioning of the heavenly bodies, the earth and the planets, generated a harmonious sound, because their distances were related to the relationships of musical intervals. According to Manly P. Hall, the distance between the planets created "sound" in tones and semitones.[39] In addition, each planet was compared to a concrete musical tone, number, color and geometric form. Each "sounded" one of the seven Greek modes (for example, Saturn "sounded" in Dorian while Jupiter "sounded" in Phrygian mode). It was established that the seven strings of the lyre correspond to the relationship between the human body and the celestial bodies. The distance between the celestial bodies "sounded" in the seven vowel letters of the Greek alphabet (Hall 1928: 82–3; see Figure 4).

---

37 Heraclitus' idea of the first taut string and his hypothesis that when the world was first created there was a strong tone is noteworthy.

38 Other expressions meaning the universal harmony/universal music: German *Sphärenharmonie*, *Sphärenmusik*, English *Music of the Spheres*, Latin *musica universalis*. The concept *universus* (Latin – global, all in one) is adapted as an expression of the world, or the Cosmos.

39 The sum of all intervals equals six whole tones, or the most perfect interval of an octave.

Figure 4. Above: sounding relationships of four elements; below: musical positioning of the heavenly bodies (the information was systematized by the author of this book – R. P.*)*

Fire : Earth	1 : 2	*perfect octave* (P8)
Fire : Air	1 : 4/3	*perfect fourth* (P4)
Fire : Water	1 : 3/2	*perfect fifth* (P5)
Air : Water	4/3 : 3/2	*whole tone*

Earth : Moon	*whole tone*	A, α (*Alpha*)
Moon : Mercury	*semitone*	E, ε (*Epsilon*)
Mercury : Venus	*semitone*	H, η (*Eta*)
Venus : Sun	*whole & semitone*	I, ι (*Iota*)
Sun : Mars	*whole tone*	O, ο (*Omicron*)
Mars : Jupiter	*semitone*	Υ, υ (*Upsilon*)
Jupiter : Saturn	*semitone*	Ω, ω (*Omega*)
Saturn : any star	*semitone*	

The idea born in Antiquity that the universe is harmoniously "linked" by a system made up of a variety of elements remained important as it was translated into the Christian worldview. The second century thinker who went by the name of Pseudo-Plutarch, declared the following maxim:

> Everything, they say, was constructed by God on the basis of musical harmony. (Mathematics and Music 2002: 1)

For example, according to the theory of the movement of celestial bodies in Antiquity, the planets were round and moved in a circular pattern (as Plato pronounced, this is "most appropriate to mind and intelligence", *Timaeus*, 34a). This thinking was connected to the interpretation of circle in the Middle Ages (in this context circle refers to a godly beginning and perfection). This concept emerged during the Renaissance too. In this epoch it was argued that perfect bodies, the planets, may move only in a perfectly circular trajectory. In baroque era music the circle symbolized the end, and group of notes (for example, phrase, motif) depicted in a circle or two half circles of necessity must be connected into unison (Lobanova 1994: 125).

A few compositions from the baroque era bear witness to the fact that in this period the concept of musical spheres was especially poignant. For example, the composer, theoretician, and organist Thomas Morley (1557/8–1602), in his treatise *Plaine and Easie Introduction to Practicall Musicke* (1597), discussed the relationship between music and universality. He created a chart that illustrated the perfect system of relationships between planets, goddesses, musical modes and Greek ideals of perfection. Robert Fludd (Robertus de Fluctibus, 1574–1637) in his treatise *De Musica Mundana* (1618) illustrated the universal concept of music in Antiquity as an example of the world's monochord. In his diagram of musical elements, he repeated the Pythagorean musical intervals and the idea of harmonious relationship between the four elements (how the proportions between the four elements were expressed in the consonances of octaves, fifths, and fourths). The German mathematician and astronomer Johannes Kepler argued that the category of numbers was universal, and at the same time an immanent part of musical structure, and that the art of sounds (music) could be expressed through numerical relationships. In his treatise *Harmonices mundi* (1619) in his theory of the mathematical movements of planets he employed a musical counterpoint, because he understood the Cosmic inheritance as a secret, occult harmony, meaning a harmony of the sounds of celestial bodies (Latin *concentus*), which God, like a conscientious organist, created according to the proportions of numbers, forming a "perfect symphony." This baroque era scholar wrote out the movement of musical planets in notes, arguing that oval trajectory of the movement of planets creates melody (according to Heher 1992: 30; also see: Kepler 1619, Bayreuthr 2004, Musik: Geschichte 1954: 166).

Athanasius Kircher (1601/2–1680) argued in his treatise called *Musurgia universalis* (1650) that there is a relationship between a micro- and macrocosmos and that its magical and rational link may be expressed through a procedure of numbers. Marina Lobanova based her work on this author's theories and raised

Figure 5. System of relationships between planets, goddesses, musical modes and Greek ideals of perfection by Thomas Morley (from Morley 1597, supplement after page 199)

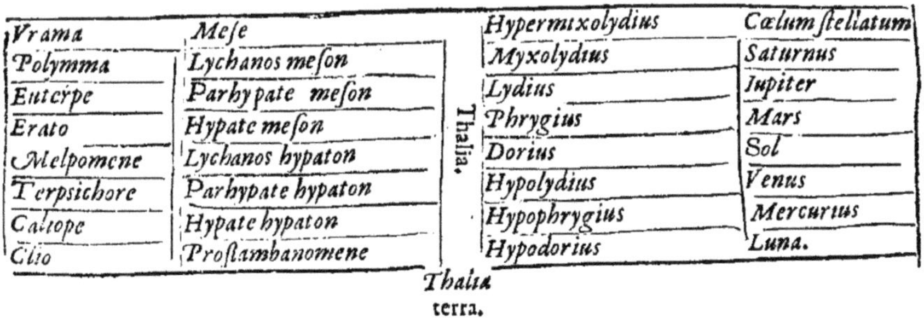

Figure 6. World's monochord in the relationship between musical intervals and the four elements as the universal concept of music in Robert Fludd's treatise De Musica Mundana *(1618)*

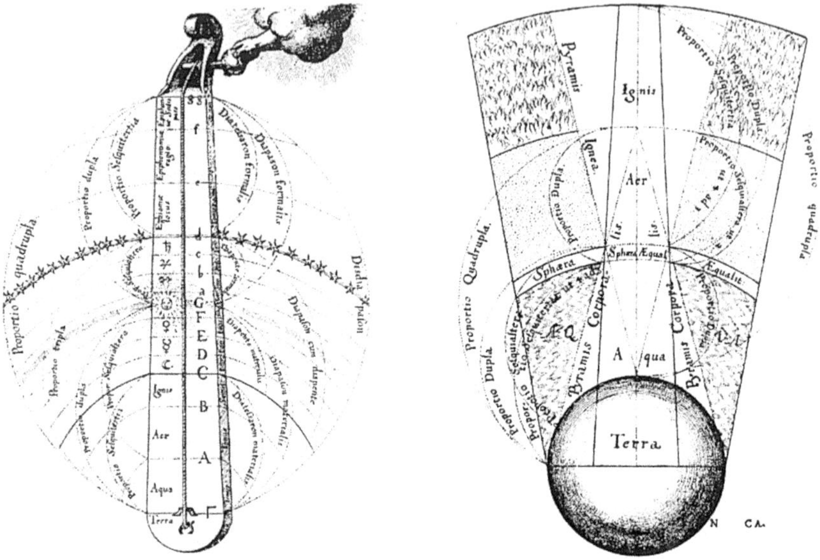

Figure 7. Equivalents of musical melodies and melodies in Kepler's Harmonices mundi *(1619, Book V, p. 207)*

HARMONICIS LIB. V. 207

mnia (infinita in potentiâ) permeantes actu : id quod aliter à me non potuit exprimi, quam per continuam seriem Notarum intermedia-

Saturnus Jupiter Mars ferè Terra

Venus Mercurius Hic locum habet etiam;

the idea that the main constant of the aesthetic idea of German baroque music was a musical inheritance that acknowledged its debt to a cosmic order and to its principles, because musical sounds capture the relationship between the micro- and macrocosmos. According to Lobanova, the theoreticians of the 17th century (Walther, Kircher, Morley, Luther and others) were convinced of the existence of a "heavenly cappella" and a musician earns his place in this cappella only if on earth he adheres to the principles of numerical harmony – the rules of sounds, as it is expressed through numbers 1, 2, 3, 4, 5, 6, and 8 (Lobanova 1994: 120–2).

1.3. The Phenomenon of *Mathesis Universalis*

The rational idea of proportions, aesthetically substantiated in the conception of the harmony of spheres as an expression of perfection, which emerged in Antiquity, was suppressed during the Middle Ages by Christian mysticism. However, with the Renaissance world outlook it came back to life as a peculiar synthesis of materiality and spirituality. It means, the rational conception of Antique beauty returned and joined the Christian world.

The issue of *quantitative* belief matured during the epoch of the Renaissance and "sin and grace were began to be measured." Symbols began to be treated as part of the work of mental power performing the function of "the systemic centre of the world" (Mažeikis 1998: 45). The human figure was established as the centre of the world – man, whom the Italian philosopher Marsilio Ficino (Marsilius Ficinus, 1433–1499) named *copula mundi*[40] became the most important point of all compositions, symmetries and perspectives concentrating the most significant relationships between micro- and macrocosmos. This concept is illustrated by Da Vinci's painting *Vitruvian Man* (c. 1487), which is regarded as the canon of proportions and which paraphrases the ideas about universal harmony expressed in the treatise *De architectura* (On Architecture, c. 15 B.C.) by the Roman writer, architect and engineer Vitruvius (80/70–c. 15 B.C.).

On the one hand, the main peculiarity of the worldview of the Renaissance was the rebirth of the philosophies of Antiquity, of Platonism, Aristotelian, stoicism or Epicureanism, as well as the formation of the conceptions of the Modern Ages. On the other hand, from the point of view of symbolic thinking and the structures of symbols created by symbolic thinking, such extremities as humanism and magic, art, science and intensified religious searches for the forms of Christianity were interrelated during that period (Mažeikis 1998: 14).

---

40 Latin *copula mundi* – the bond and knot of the world.

The philosophy of the Renaissance is characterized by its polyhedral nature – we may mention symbolic philosophies of love, *studia humanitatis* (philological humanism), *studia divinitatis* (Christian humanism) or occult philosophies – trends of hermeticism, alchemy, and the Christian Kabbalah. Alongside these, there was the theory of *mathesis universalis*[41] in which a general conception of mathematics as that of real beauty ranged from the physical to sacralized, mystified dimensions. Catholic Cardinal Nicholas of Cusa (1401–1464),[42] the author of the theory *mathesis universalis*, raised the question of *the possibility of knowing* God, rather than the question *about* God, which was possible only when man compared or measured the known and the unknown. Therefore, one had to choose the science of mathematics as one of the measures of knowledge. Gintautas Mažeikis notes that such mathematical conclusions as the ability to measure the work of God and express it in terms of numbers resonates with the truths of the Kabbalah, because:

> The letters of the Hebrew alphabet have their own numerical equivalents, and therefore God's names, and the names of angels, can be expressed in terms of numbers. (Mažeikis 1998: 142)[43]

Mathematics was a part of theology to Nicholas of Cusa, which raised reason as a measure of all earthly things and God's prototype. Therefore, rephrasing antique ideas in the Christian environment is obvious in Nicholas of Cusa's theory, because numbers to this Renaissance thinker were also a symbolic prototype of things, a creating rudiment of intellectual activity. In his didactic sermon *Tota pulchra es, amica mea* (On Beauty, 1443) he wrote the following:

> [E]very number is present in an enfolded way, and just as in number all proportion and all intermediateness are present enfoldedly, and just as in proportion all harmony and order and concordance [are enfolded]: so too, for this reason, [there is enfolded in oneness] all beauty, which shines forth in the ordering and the proportion and the concordance. Hence, when we say that God is One, this One is Supersubstantial Oneness itself, which is also Beauty, enfolding in itself all things beautiful.[44]

---

41 *Mathesis universalis* is mystic, sacral mathematics that seeks to get to know the Almighty God through different numerical and geometrical symbols, to explain his manifestations in nature and human activity.

42 Nicholas of Cusa also referred to as Nicholas of Kues, Nicolaus Cusanus; real name – Nikolaus Chrifftz or Krebs.

43 Mažeikis' statement was based on Gershom Scholem's *Origins of the Kabbalah* (Princeton University Press, 1990).

44 Quotation cited in *Nicholas of Cusa's Didactic Sermons: A Selection*, transl. Jasper Hopkins, Arthur J. Banning Press, 2008, p. 174.

He regarded music as a participant in the process of the creation of the world. The passage that was presented by Cusa in his *De Docta Ignorantia* (On Learned Ignorance, 1440, Book II, Chapter 13) confirms the continuation and significance of quadrivium tradition in the Renaissance theory of *mathesis universalis*:

> [I]n creating the world, God used arithmetic, geometry, music, and likewise astronomy. [...] For through arithmetic God united things. Through geometry He shaped them, in order that they would thereby attain firmness, stability, and mobility in accordance with their conditions. Through music He proportioned things in such way that there is not more earth in earth than water in water, air in air, and fire in fire, so that no one element is altogether reducible to another. As a result, it happens that the world-machine cannot perish.[45]

1.4. Expression of Numerical Proportions and Progressions

Nicholas of Cusa's considerations testify to the synthesis of the constructive antique and sacral Christian attitudes that took place on the plane of the philosophical world outlook of the Renaissance. On a practical plane, the admiration for pure proportions of numbers was still alive, for example, as perspective in art, striving for symmetry in architecture, the rudiments of the visual arts in music rendering a composition visually perfect, graphic, and even architectural forms. Numerical proportions amply mentioned in the treatises on music during the Renaissance show that the mathematical aspect in music has survived as the main factor of beauty that is only slightly affected by the conception of sacrality. Though in his *Proportionale musices* (Proportions of Music, c. 1472–5) Johannes Tinctoris (c. 1435–1511) stated that proportions existed in everything, because "it was God who created them." The constructive element was obvious in Tinctoris' description of musical proportions and their significance:

> [M]any other famous composers whom I admire, while they compose with much subtlety and ingenuity and with incomprehensible sweetness, are either wholly ignorant of musical proportions or indicate incorrectly the few that they know. I do not doubt that this results from a defect in arithmetic, a science without which no one becomes eminent, even in music, for from its innermost parts all proportion is derived. (Source readings 1965: 5).

Tinctoris and other theoreticians of the Renaissance, following the science of Boethius, classified the proportions applied to the creation of music into five types characterized by the diminution process. Having retrogradely overturned

45 The translation of Cusa's passage cited in: *Complete Philosophical and Theological Treatises of Nicholas of Cusa*, Vol. 1, transl. Jasper Hopkins, Arthur J. Banning Press, 2001, p. 99.

these formulas, augmented variants were derived, which were used with the prefix *sub-* (for example, *proportio subtripla* 1 : 3):

- The first type formula n : 1 (2 : 1, 3 : 1, 4 : 1 and so on) is made up of a *genus multiplex* group. These numerical formulas are given the names: *dupla, tripla, quadrupla* and so on.
- The second type n + 1 : n (2 : 1, 3 : 2, 4 : 3 and so on) group was named *genus superparticularis*, and each of its formula's names began with the prefix *sesqui-*. For example, *sesquialtera, sesquitripla, sesquiquarta* and so on.
- The third type n + 2 : n, n + 3 : n, n + 4 : n and so on, formulas were made of the *genus superpartiens* group. The name of these proportions ended with the number, "n" and the word *super...partiente* in its middle denoted the quantity added. For example, *proportio supertripartiente quintas* is expressed in numerals that appeared as follows:
  n + 3 : n = 5 + 3 : 5 = 8 : 5.
- The fourth type n x m + 1 : n – *genus multiplex superparticulare* proportions group. For example, *proportio tripla sesquiquarta* was expressed in the formula:
  n x 3 + 1 : n = 4 x 3 + 1 : 4 = 13 : 4.
- The fifth type n x m + 2 : n, n x m + 3 : n, n x m + 4 : n and so on – *genus multiplex superpartiens* group. For example, *proportio quadrupla supertripartiente quintas* was expressed in the numbers:
  n x m + 3 : n = 5 x 4 + 3 : 5 = 23 : 5.
  The fourth and fifth types in practice were applied only very rarely because of their complexity and unwieldy nature.

The theory of complex proportions in the Renaissance was expanded and made more diverse by the formation of polytempos and polyrhythm. The example of mensural notation – Guillaume Dufay's (c. 1397–1474) motet *Alma redemptoris mater* (the date is unknown) – is indicated as one of the first examples of the polyrhythm phenomenon. There the smallest rhythmic value *minima* is divided according to the formula 4 : 3. By comparing different mensures between separate voices, a polyrhythmic pattern is obtained. The rhythmic values of different voices in Part 2 of *Agnus Dei* of Josquin des Prez's (c. 1450–1521) Mass *L'homme armé* (supposed to have been composed c. 1489–95) create a polyphony of three different movements.

The Fibonacci Progression and the Golden Ratio Phenomenon

The mathematical proportion described in the Renaissance played a significant role in creating music. This is the so-called infinite Fibonacci sequence, which purifies the Golden Ratio phenomenon ever more, and whose major principle is that each number is the sum of the two numbers before it:[46]

0–1 (1 + 0) – 1 (1 + 1) – 2 (2 + 1) – 3 (3 + 2) – 5 (5 + 3) – 8–13 – 21–34 – 55–89 and so on

---

46 In one of his works *Il liber abbaci* (1202) the mathematician Fibonacci (also referred to as Leonardo Pisano, Leonardo da Pisa, 1180–1250; the name Fibonacci is derived from *filius Bonassi*, Latin – son of Bonassi) propagated the Arabic tradition of numerology and criticized the cumbersomeness of the Roman one.
Fibonacci's famous sequence can be considered to be a numerical expression of the principle of nature/evolution thereby the mathematician of the 13th century represented the process of reproduction of rabbits: Fibonacci made calculations of how many young one pair of rabbits will produce per year, having in mind the fact that all the young will survive and reproduce further. Having summed the number of rabbits of each month, the progression of numbers 1–1–2–3–5–8–13–21... was obtained. It is the so-called recurrent sequence of the second order (denoted $F_1, F_2, \ldots, F_n$), based on the fact that its term $n + 2$ is equal to the sum of the two preceding terms (n and $n + 1$). Summing determined the formation of the infinite sequence, and the formula of progression was written down $F_{n+2} = F_{n+1} + F_n$, where $n \geq 0$.
Commenting on the Fibonacci calculations, the musicologist Newman W. Powel compiled a table in which horizontals and verticals, as well as the sum of the horizontals or verticals, formed the Fibonacci sequence. The calculations were started in December, when a pair of rabbits had no young yet; the month of January was devoted to mating (each of these months was denoted by number 1); the process of the reproduction of rabbits started in February (the scheme reproduced from Powel 1979).

		December	January	February	March	April	May	June	July	August	September	October	November	December	January	Total
0		1	0	1	1	2	3	5	8	13	21	34	55	89	144	377
1	mating		1	0	1	1	2	3	5	8	13	21	34	55	89	233
2	offspring			1	0	1	1	2	3	5	8	13	21	34	55	144
3	offspring				1	0	1	1	2	3	5	8	13	21	34	89
4	offspring					1	0	1	1	2	3	5	8	13	21	55
5	offspring						1	0	1	1	2	3	5	8	13	34
6	offspring							1	0	1	1	2	3	5	8	21
7	offspring								1	0	1	1	2	3	5	13
8	offspring									1	0	1	1	2	3	8
9	offspring										1	0	1	1	2	5
10	offspring											1	0	1	1	3
11	offspring												1	0	1	2
12	offspring													1	0	1
13	offspring														1	1
Total		1	1	2	3	5	8	13	21	34	55	89	144	233	377	*986*

Historically, the source of this number sequence is attributed to Euclid's ideas in Antiquity, c. 300, about a certain relationship, which he called "the special proportion." However, in Euclid's treatise *Στοιχεῖα/Stoicheia* (Elements, date of publication c. 300 B.C.) this proportion was analyzed not as a particular row of numbers, but as a

The rules of the Fibonacci sequence may be applied to any sequence of numbers in which each new number is the sum of the two previous numbers. For a long time, according to this principle, a variety of numeric progressions were calculated, for example the Lucas series or *Série Évangélique*:[47]

Série Évangélique
2–5 – 7 (2 + 5) – 12 (5 + 7) – 19 (7 + 12) – 31–50 – 81–131 and so on

Lucas progression
2–1 – 3 (2 + 1) – 4 (1 + 3) – 7 (3 + 4) – 11–18 – 29–47 – 76–123 and so on

The Lucas and *Série Évangélique* progressions were based on the text of the Holy Bible that was interpreted using symbolic numbers. For example, the first numbers of *Série Évangélique* were taken from the Gospel according to John (6,9–13; italics are mine – *R. P.*):[48]

---

specific geometric manifestation. To arrive at his proportions, he relied on lines and interrelationships between geometric constructions. For example, according to the Greek mathematician, the line *c* divided into two parts, and the relationship between the smaller *b* and the larger *a* is equal to the relationship between larger section *a* and all line *c*, that is:

$$b : a = a : (a + b) = a : c$$

At the same time, Euclid analyzed typical relationships of the constructions of the rectangle, square, the regular pentagon, and dodecahedron (one of Plato's five perfect polyhedra, which has 12 flat faces made up of regular pentagon forms) or the icosahedron (another one of Plato's bodies, which has 20 flat faces in a regular triangle form). Ruth Tatlow argues that using a similar principle based on particular numbers in the 13th century, Leonardo Pisano did not notice links between his discovery and Euclid's "special proportions." He did not denote that his Fibonacci numbers have "special relationships" (Tatlow 2006: 77). This would explain how it was not a novelty in the 13th century to use addition in number sequences.

47 In the 19th century the Lucas series was named after the French mathematician Édouard Lucas (1842–1891). The French title *Serie Évangélique* can be found in Georgeso Arnoux's book *Musique Platonicienne Ame du monde* (Paris: Dervy-Livres, 1960, pp. 222–3). However, until they were given names in the 19th–20th centuries these progressions were known as the expression of the tenth Neo-Pythagorean proportion. They were described by the philosopher of late Antiquity Nicomachus (Νικόμαχος, c. 60–c. 120) in his treatise *Arithmetike eisagoge* (Introduction to Arithmetic, date of publication unknown), where he argues that the tenth proportion is created when the relationship between two smaller numbers equals a larger relationship. For example, 3 (a) with 5 (b) and with 8 (c).

48 Newman W. Powel showed in his calculations how the Lucas and *Serie Évangélique* numerical progressions were derived from the Fibonacci sequence, using the formulas F_{n+1} or F_{n-1}. The derivation of sequences is displayed in the table below (made accord-

> There is a boy here who has *five* barley loaves and *two* fish, [...] Jesus said, "Have the people sit down." Now there was much grass in the place. So, the men sat down, about *five* thousand in number. [...] And when they had eaten their fill, he told his disciples, "Gather up the leftover fragments, that nothing may be lost." So they gathered them up and filled *twelve* baskets with fragments from the *five* barley loaves left by those who had eaten.

Though what made the numerical series exclusive, which attracted the attention of thinkers and creators of different epochs? Perhaps the most important feature of the Fibonacci progression is not the individual numbers, but the fact that by means of the ratio between them it is possible to come maximally close to the Golden Ratio formula,[49] which expresses the laws of nature. In 1509, in his treatise *De Divina Proportione,* Luca Pacioli looked at this formula emotionally and called it "the divine proportion", while the term *sectio divina* was proposed in the beginning of the 18th century by the astronomer Kepler, to whom the following saying is attributed:

> Geometry has two great treasures, [...] one is the Theorem of Pythagoras; the other, the division of a line into extreme and mean ratio. The first we may compare to a measure of gold; the second we may name a precious jewel. (Rothstein 1995: 162)

The conception of the Golden Ratio appeared in 1835 when the German mathematician Martin Ohm (1792–1872) named the so-called "constant proportion" (German *Stetige Proportion*) the *Goldener Schnitt*. Incidentally, the very sequence of numbers was named after Fibonacci and was related to the phenomenon of the

---

ing to Powel 1979: 231). Lucas numbers (marked as L_n, $n \geq 1$) are obtained summing up F_{n-1} and F_{n+1}. While by adding F_{n+1} to Lucas numbers we get the sequence of *Serie Évangélique* (marked as E_n, $n \geq 1$).

F_{n-1}	0	1	1	2	3	5	8	13	21	34	55	89	144
$+ F_{n+1}$	1	2	3	5	8	13	21	34	55	89	144	233	377
$= L_n$	1	3	4	7	11	18	29	47	76	123	199	322	521
$+ F_{n+1}$	1	2	3	5	8	13	21	34	55	89	144	233	377
$= E_n$	2	5	7	12	19	31	50	81	131	212	343	555	898

49 Golden Ratio – Latin *sectio aurea, sectio divina*, German *Goldener Schnitt*. The Golden Ratio is expressed in terms of the formula *n x 0.618*. The fraction 0.618 was obtained having evaluated the ratios between the Fibonacci numbers. It became more accurate the more distant the pairs of Fibonacci numbers calculated, because 3 : 5 = 0.6; 5 : 8 = 0.625; 8 : 13 = 0.61538461...; 13 : 21 = 0.61904761...; 21 : 34 = 0.61764705... and so on. The exact mathematical value of the Golden Ratio is $\frac{\sqrt{5}-1}{2}$.

Golden Ratio only in the 19th century,[50] when the progression of numbers was noticed to acquire an ever more exact expression of the Golden Ratio.

The Golden Ratio became the benchmark and the goal, and the guarantee, of perfect art. According to the measurements of the Golden Ratio, buildings were constructed, parks were planned, compositional details were applied to painting, and poetic stanzas were constructed. In music the formula of godly beauty symbolized a perfectly formed composition. Often the climax, the most important part of any composition, matched the point of the Golden Ratio. Sounds were organized and regulated according to Fibonacci principles of numeric progression, and so on. For example, while analyzing the structure of two vocal ballads from the Middle Ages, *Dame, se vous m'estes lonteinne* and *Je ne croy pas c'onques a creature,* by the composer Guillaume de Machaut (c. 1300–1377), Pozzi Escot shows that the limits of the structural parts adhere to the Golden Ratio (Escot 1999: 43, 51).

During the Romantic period, which Cholopov called the culmination of the antagonism between music and mathematics (Cholopov 1982: 78), where anti-rationalism was especially strong and where it would seem that emotionality was especially important, Fibonacci numbers and the Golden Ratio prevail nonetheless. This is also true of Frédéric Chopin's preludes (more about preludes see: Kirk 2001, Escot 1999), Ludwig van Beethoven's Fifth symphony in the structure of the first part (see Madden 1999 & 2005) or in the structures of the compositions of Claude Debussy (for example see: Howat 1983, Soussidko 2002). According to Allan W. Atlas, structural and harmonic changes are important in the duets, arias, and orchestral interludes of Giacomo Puccini's opera *La bohème* and reflect the relationship of the Golden Ratio (more see Atlas 2003).

An analysis of Chopin's Prelude No. 1 in C major, Op. 28, shows several instances in which Finobacci numbers 8, 13, 21, and 34 make an appearance. This miniature composed for piano according to the form of the period is made of two phrases consisting of 8 and 16 measures, in addition to the establishment of C major in a 10 measure coda. For example, in the 8th measure the lowest tone of the prelude is reached, the tone G_1 (contra-octave *sol*), in the 13th measure the first meaningful step towards harmony is taken, a modulation to D minor, from

50 In 1857, the Prince of Italy, the mathematician and historian Baldassarre Boncompagni (1821–1894), published a medieval treatise; in 1878, the French mathematician Édouard Lucas, having become acquainted with the treatise, named the said sequence the Fibonacci Sequence. Basing herself on that, Ruth Tatlow believes that the statement discussed by Boethius about the numbers called Fibonacci is incorrect (Tatlow 2006: 74, 77–8).

Figure 8. Chopin, Prelude No. 1 C major, Op. 28. Manifestation of Fibonacci numbers in the structural points of the composition

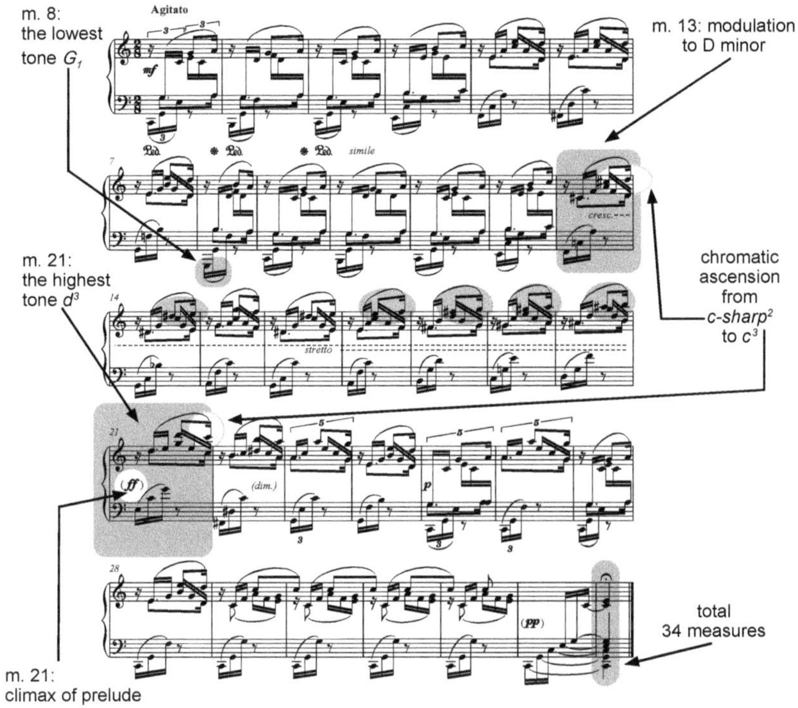

this measure the diatonic picture of the composition is changed by the chromatic melody rising from *c-sharp*^{2} to c^3 and then deepened by relief of a modulation. From there in the 21st measure the highest tone of the prelude sounds, d^3, and then the climax. The chromatic movement of the melody is changed by a diatonic movement downwards. The prelude is made up of 34 measures (see Figure 8).

Debussy's work can be described as a medley of impressionistic tonal images and emotion, which is frozen by movement and a moment. This French composer referred to music as "the silence between sounds." However, the foundation principles of his compositions are dictated by a constructive logic. He has been quoted as stating: "Music is the arithmetic of sounds as optics is the geometry of light."[51]

---

51 The composer's quotation was published in Dean Keith Simonton's *Greatness: Who Makes History and Why* (Guilford Press, 1994, p. 110).

Figure 9. The symmetrical groups of measures in the piano preludes by Debussy (reproduced from Soussidko 2002: 9–12)

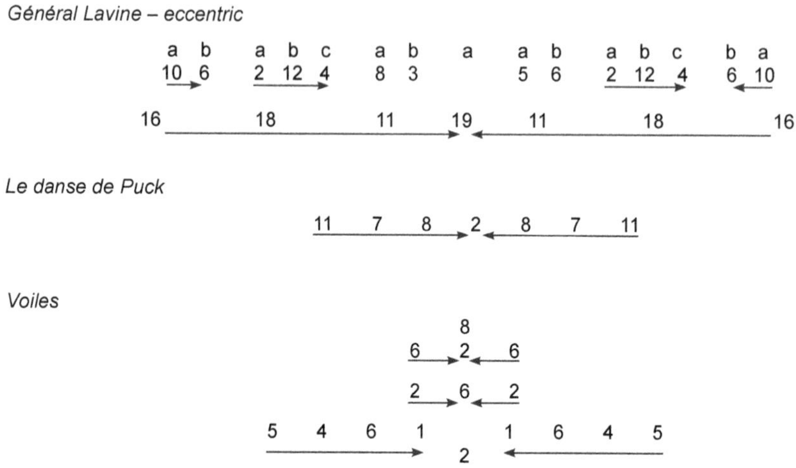

For example, Roy Howat, while analyzing Debussy's *Dialogue du vent et de la mer*, the third part from his triptych *La mer* (1903–5), or the first piece *Reflets dans l'eau* from his cycle for piano *Images* (1904–5), notes that the same wave in their structure can be seen in Katsushika Hokusai's colored woodcut *The Great Wave off Kanagawa* (1831), which illustrates the phenomenon of the Golden Ratio (Howat 1983: 23–9, 93–109; also see *Golden Ratio* 1988). In the structure for his preludes for piano *Général Lavine – eccentric, Le danse de Puck* or *Voiles* (preludes composed in 1910–1913) it is possible to find the mirror image, or symmetry models, which express themselves in the proportions of measures (see Figure 9 based on Soussidko 2002: 9–12).

1.5. *Ars Combinatoria* and the Constructivism of Music

Another possibility of numerical manipulations was *ars combinatorial*,[52] made perfect in the practical space of Renaissance music. The combinatory art of composing music is not related to any particular musical genre, style, and type of form or compositional technique, because *ars combinatoria* manipulations are universal permutations (limited rearrangements of elements) and combinations (replacing some elements by others, a huge number of variants) applied by manipulating with various parameters

52 The concept *ars combinatoria* is found in the treatises by Mersenne (*Harmonie universelle*, 1636), Kircher (*Musurgia universalis*, 1650) and Leibniz (*Ars combinatoria*, 1666); according to Gerver & Lebedeva 1999: 30–2.

of a musical composition – pitches, rhythmic units, segments of polyphonic lines, the *cantus firmus* structure, etc. The adaptation of transformations (grouping, rotation and combinations) in music can be compared to a literary anagram[53] as different combinations of sounds of a melody can be formed on a similar principle.

Most probably the composers were intrigued by the *ars combinatoria* due to the infinite possibilities of manipulations. For example, a melody of eight tones has all in all as many as 40,320 variants of transpositions (they are obtained by multiplying all eight numbers 1 x 2 x 3 x 4 x 5 x 6 x 7 x 8).

The Russian musicologist Nikolai Tarasevich, when trying to understand how the *ars combinatoria* phenomenon (as well as other possibilities of mathematical nature of composing music) drew the composers' attention, reduced the inclination of the Renaissance composers to perform complicated combinatory actions in a musical composition to the human *lusus* (Latin – game), to the playing element peculiar to the aesthetic and artistic activity of that period when "the number functioned not as a number, but as an expression of a game" (Tarasevich 1992: 44). This is in line with the Renaissance conception of man's humanness (*homo humanus*), which, according to Martin Heidegger (1889–1976):

> ... was opposed to *homo barbarus*. *Homo humanus* here means the Romans who exalted and honored Roman *virtus* through the "embodiment" of the *παιδεία* [education] taken over from the Greeks.[54]

The philosophers of the Renaissance epoch often referred to *homo sapiens* as *homo ludens* (Latin – the playing man). Johan Huizinga (1872–1945) who published the book *Homo Ludens* (*The Playing Man*) in 1938 thoroughly investigated the concept of "the playing man" as the concept of the theory of the game. Hermann Hesse (1877–1962) wrote the following about the rule of the game:

> [T]he Glass Bead Game player plays like the organist on an organ. And this organ has attained an almost unimaginable perfection; its manuals and pedals range over the entire intellectual cosmos; its stops are almost beyond number. Theoretically this instrument is capable of reproducing in the Game the entire intellectual content of the universe.[55]

---

53 An anagram is the permutation of letters or syllables when another word or a meaningful phrase is formed. For example, striking possibilities of the anagram are revealed by an example in the English language when having transposed the letters of the phrase "twelve plus one" we receive "eleven plus two."

54 The statement was published in Heidegger's *Letter on "Humanism"* in 1947; cited in Martin Heidegger's *Pathmarks*, Frank A. Capuzzi (transl.), William McNeill (ed.), Cambridge University Press, 1998, p. 244.

55 Cited in Hesse's *Magister Ludi* (The Glass Bead Game, Richard and Clara Winston (transl.), New York: Bantam Books, 1969, p. 6).

Figure 10. Obrecht, Missa Grecorum. *Original* cantus firmus *in* Kyrie, *mm. 1–24, and its permutation in* Credo, *mm. 1–40*

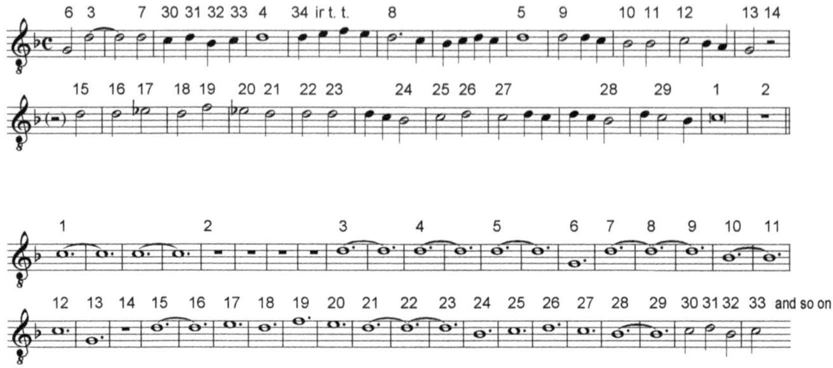

An interesting and sophisticated game, namely the application of combinatorics to a Renaissance era musical composition, was undertaken by Jacob Obrecht (1457/8–1505) in the score of his *Missa Grecorum* (1503). The technique of rearranging the unique tones of the *cantus firmus* would be difficult to know without knowing its "secret." For example, at the beginning of the *Credo* part, it sounds as though accidental *cantus firmus* tones are being used. That is because the *cantus firmus* in the tenor part is transformed unrecognizably. In this section the composer applied an especially refined technique of combinatorics – he recomposed the original melody in the tenor score, having first chosen and written out the longest tones and rests of *cantus firmus*, then the shorter tones to the very shortest (see Figure 10).

Discussions of the art of musical combinatoric was especially active in the 16th and 17th centuries. The theoreticians of that era understood music as the result of the calculations of combinatorics. They offered a variety of experiments. For example, the so-called musical generating machines. One of the first important works dedicated to musical combinatorics was Marin Mersenne's (1588–1648) *Harmonie universelle* (1636). In his section on practical applications of musical combinatorics the scholar presented all possible (720) hexachord compounds.

Combinatory processes were analyzed in music theory soon after Mersenne's work was published. In his treatise *Musurgia Universalis* (Universal Music-Making, 1650) Kircher argues that musical composition is like a combinatory equation (Gardner 1974: 132). He writes about a mechanical object used to compose music.[56] According to Kircher, the so-called idea of *arca musarithmica*, the arithmetic relationship of numbers was used to create rhythm, relationships and changes of tempo, and tonalities in music (Klotz 2006: 38).

In 1739 the general bass, or figurative bass, was described as a process of combinatory music, according to Mizler in the periodical *Neu eröffnete musikalische Bibliothek*. Joseph Riepel (1709–1782) in his treatise *Gründliche Erklärung der Tonordnung insbesondere* (1757) discusses *ars combinatoria* in the practice of artistic music and gives examples on how to create tonal combinations of two, three, or four tones.[57]

In the treatises of the 18th and 19th centuries, there are many suggestions on how to compose so-called dice music, the most typical representation of numerology in classical music. In the treatises at least twenty suggestions are given on how to compose music with different numbers, because "anyone who knows dice with numbers and can copy music is able to compose so many of the aforementioned

---

56 The first "machine" to compose music (called *musarithmica mirifica*) was constructed in the 17th century. It was modeled by Samuel Pepy based on Kircher's treatise.

57 In this treatise, Riepel presented playful names for diatonic scales. The names described the relationships between tones and accurately reflected the hierarchy of tones in the tradition of Baroque music (Riepel 1757: 4):

C – "The Landlord" (*Meyer*)
G – "The Main Servant" (*Oberknecht*)
a – "The Main Maid" (*Obermagd*)
e – "The Second Maid" (*Untermagd*)
F – "The Day Laborer" (*Taglohner*)
d – "The Hired Laborer" (*Unterläufferin*)
b – "Black Margaret" (*Schwarze Gredel* – the nickname of the Queen of Sweden, who had a swarthy look to her)

small pieces with one or two dice."[58] This quote was taken from one of the most influential treatises on music composed by dice written by a student of Johann Sebastian Bach, Johann Philip Kirnberger (1721–1783). Musicologists Ruth Tatlow and Paul Griffiths confirmed that Kirnberger's surviving polonaises were based on his compositional method with dice (Tatlow & Griffiths 2016). In 1783 another book by this theoretician was published. This book was dedicated to the composition of symphonies and other musical forms by using the same method.

In 1779 the Austrian composer Maximilian Stadler (1748–1833) published charts and examples of musical measures, which anyone could use to compose minuets and trios. Kirnberger's method of composing music by using dice can be found in treatises from the end of the 18th century. The authorship of these works is attributed to Haydn and Mozart. For example, a treatise titled *Musikalisches Würfelspiel* (published in 1792, Amsterdam) was one of the most popular treatises on the subject, and provided instructions on how to compose German waltzes using two dice. This treatise is often attributed to Mozart, because he enjoyed mathematical puzzles. His manuscripts show the use of permutation technique in musical practice. Also well known is the 1768 publication of Herman François de Lange's treatise *Le toton harmonique*, which offers a method based on nine-sided top. A version of this method can be found in Carl Philipp Emanuel Bach's article "Einfall, einen doppelten Contrapunct in der Octave von sechs Tacten zu machen, ohne die Regeln davon zu wissen" (A Method for Making Six Bars of Double Counterpoint at the Octave Without Knowing the Rules, c. 1757) (Tatlow & Griffiths 2016).

According to the rules described in *Musikalisches Würfelspiel*, a 16-measure waltz could be composed by tossing two dice 16 times. The system of composition was made up of 176 numbered examples of short one-bar music phrases. Out of these, using two number charts each for 8 columns, 16 bars were selected (each column was written with 11 numbers from 2 to 12, which were achieved by throwing dice, from the smallest number of 2 [1 + 1] to the largest 12 [6 + 6]). The first chart and first eight tosses were used to compose the first eight measures of the waltz. The second chart and eight tosses determined the next eight measures of the composition. The generalizing principle was based on the logic of classi-

58 Original quote in German from Johann Philip Kirnberger's treatise *Der allezeit fertige Polonoisen- und Menuettencomponist* (The Ever-Ready Polonaise and Minuet Composer, Berlin, 1757, p. 2):

> Ein jeder der nur Würfel mid Zahlen kennet, und Noten abschreiben kann, ist fähig, sich daraus so viele der genannten kleinen Stücke, vermittelst eines oder zweener Würfel zu componiren.

cal functional harmony – the beginning, T (tonic), the first eight-measure end modulating to D (dominant), and the end returns to T. By using this system of composing music by dice, it was possible to create a thousand versions of a waltz.

Therefore, the concept of "musical mathematics" could be applied to classical music. This period extended the rational worldview of Antiquity and its typical forms of architectonic proportions into a widely understood symmetry. All that was shown through the harmony of mathematically balanced elements of form. For example, Leopold Brauneiss revealed the influence of symmetry on the structural organization of Mozart's composition *Sonata facile* (Sonata in C major, KV 545) and displayed the relationship of the numbers of bars between the three parts (the number of measures in part 1 is 73, part 2–74, part 3–73) (Brauneiss 1995: 86–90).

2. Semantic Interpretation of the Interaction between Music and Mathematics: Mystic Middle Ages and the Sacral Baroque

The Middle Ages is the epoch that divided a dynamic Antiquity and the era of Renaissance, referred to as *medium aevum*, and is characterized by the humanists, or the 19th century aesthetes, with the concept of having the negative shade "centuries of darkness" or the period of "stagnation." By its dominating theocentric perception of the world, this epoch clearly contrasted with antique anthropocentrism, which later would be revived by the Renaissance world outlook. A surprisingly sudden establishment of a religious way of thinking in all spheres – in arts, sciences, and especially in everyday life – is related to a new, stable and firm faith – Christianity.

Nonetheless, the medieval perception of the world was based on the relics of antique culture. The medieval thinkers did not reject the ideas of the previous epoch and often based themselves on them. For example, the rational conception of beauty formed in ancient Greece expressed by the order of numbers, relations of numerical proportions, also formed the material basis of the medieval world based on sacral thinking. The same mathematical proportions were perceived as attributes of transcendental perfection and the antique *techne*[59] conception was taken over in the Middle Ages as the ability to correctly and skillfully create or understand some craft well. For example, Boethius, when writing about the relation between arithmetic and the science of music, based himself on the observations of the philosopher of the Late Antiquity Nicomachus (Νικόμαχος, c. 60–c. 120).

In the Middle Ages, however, antique ideas were treated through the prism of the Christian faith, the aesthetic thought was unconditionally subject to theology. This is illustrated by the spread of antique *quadrivium* – the medieval world classification of seven sciences (*septem artes liberales*) included also branches of quadrivium and was interpreted as necessary preparatory sciences studied before going deep into the mother of all sciences – theology. The impact of the Christian faith is obvious on a creative plane too. The creative power was perceived as sacred empowering or a gift granting the earthly subject – man to carry out God's mission on Earth – to glorify the Almighty God and His deeds.

---

59 Greek *τέχνη* – art.

2.1. Semantics of the Kabbalah in Music

Constructive manipulations with numbers, which intrigued one for their mathematical subtleties and which became a separate science – the Kabbalah – were related to the proof of divinity in medieval Christian culture. The expression of mysticism – the formation of various geometric shapes, for example, magic squares of numbers, spread in the world of numbers. The principle of their structure gave them that mysterious meaning: when summing different numbers horizontally, vertically or diagonally (each number is found only once in the square) the same result was obtained.[60]

Of great importance in this sense was Heinrich Cornelius Agrippa's (1486–1535) treatise *De occulta philosophia libri tres* (Three Books of Occult Philosophy, first published in 1531), which described seven magic squares of numbers of 3–9 rows related to the planets. For example, number 152 "radiating" the sacralized meaning was found in the Christian tradition – the numerical symbol of Maria (MARIA/MAPIA = 152) through the magic square could have been related to the name of Christ, because the sum of the Greek name ΙΗΣΟΥΣ (Greek – Jesus) is 888 (Ι – 10, Η – 8, Σ – 200, Ο – 70, Υ – 400, Σ – 200).[61]

---

60 The compilers of the squares of numbers were intrigued by other possibilities of these models that required special logical manipulations. For example, how to form a pandiagonal magic square (e.g., for each microsquare that is formed inside the macrosquare, the elements of both diagonals of each microsquare add up to the same number. For example, if a square is divided into four squares as shown in the figure below, all small main diagonals add up to 130; if the square is divided into nine small squares, this sum is different for every microsquare, but both main diagonals of each add up to the same number), composite magic square (composed of separate magic squares), or concentric/framed magic square (principles of the magic number remain after the top and bottom rows and the side columns are removed from the macrosquare).

17	46	12	55	54	9	47	20
16	51	21	42	43	24	50	13
53	10	48	19	18	45	11	56
44	23	49	14	15	52	22	41
25	64	2	39	62	27	37	4
8	33	31	58	35	6	60	29
63	26	40	1	28	61	3	38
34	7	57	32	5	36	30	59

71	66	67	20	25	24	29	34	33
64	68	72	27	23	19	36	32	28
69	70	65	22	21	26	31	30	35
8	3	4	40	39	44	74	79	78
1	5	9	45	41	37	81	77	73
6	7	2	38	43	42	76	75	80
47	54	49	56	63	58	11	16	15
52	50	48	61	59	57	18	14	10
51	46	53	60	55	62	13	12	17

64	4	9	54	63	3	10	63
60	15	16	47	48	49	20	5
7	44	22	42	41	25	21	58
51	33	37	29	30	28	38	14
6	32	34	35	36	31	27	59
8	26	40	24	23	43	39	57
52	45	46	18	17	19	50	13
12	61	56	11	2	62	55	1

61 For more about the equivalents of letters and numbers see section "Codes of the Numerical Alphabets in Music" in this book, pp. 60–68.

Figure 11. The magic square with numerical symbols of Maria and Christ (152 and 888)

1	5	2	8
5	2	1	8
2	1	5	8
8	8	8	8

Other magical manipulations of numbers were also mentioned – these were the magical triangle, the hexagon, stars, and other figures. For example, while organizing one variant of the magical triangle, its left side is made from a sequence of natural numbers. On the right side numbers are obtained by adding an integer by 1 greater each time; thus we obtain the so-called triangular numbers. The so-called magical hexagon is made from a part of a regular hexagonal lattice. The numbers are written in; the sum of the numbers in any direction is always the magic constant. The polygonal star is made magic through the sum of the numbers written at its top and on its sides. Also, arithmetic counting was admired. For example, certain mathematical actions with the first nine numbers allow a sequential retrograde:

$123456789 \times 8 + 9 = 987654321$

The composers of the Baroque era were acquainted with the techniques of the Kabbalah and applied them practically to their compositions. The musicologist Piet Kee argues that the passacaglia and chaconne ostinato structure established the so-called magic triangle numbers, which belonged to the old triangle group. For example, Bach's chorale *Wir glauben all an einen Gott*, BWV 680, the theoretician designated as an example of number 6 – one of the members of the magic triangle, while in the structure of Georg Böhm's (1661–1733) compositions *Chaconne, Präludium, Fuge und Postludum* he analyzed the influence of the number 10 (Kee 1988: 233). It is also apparent that these numbers are significant in Christian numerology.

Figure 12. Magic number triangle, number circle (the sum of numbers is 13) and number tree. Two magic hexagons with "magic" constants 38 and 244. The examples of magic stars: the hexagonal star with constant 26, the heptagon star with constant 30 and the octagonal star with constant 34 (systematized by the author of this book – R. P.*)*

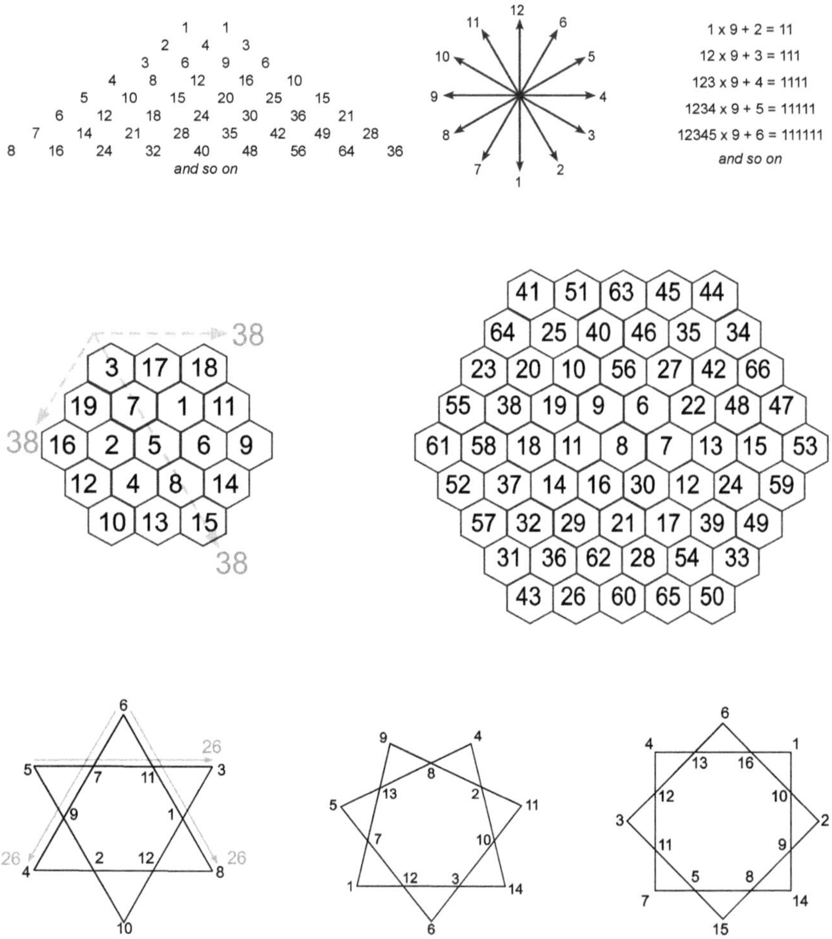

Figure 13. Seven magic squares related to the planets according to Agrippa (diagrams based on Agrippa 1533: CXLIX–CLIII)

Saturn =15		
4	9	2
3	5	7
8	1	6

Jupiter = 34			
4	14	15	1
9	7	6	12
5	11	10	8
16	2	3	13

Mars = 65				
11	24	7	20	3
4	12	25	8	16
17	5	13	21	9
10	18	1	14	22
23	6	19	2	15

Sun = 111					
6	32	3	34	35	1
7	11	27	28	8	30
19	14	16	15	23	24
18	20	22	21	17	13
25	29	10	9	26	12
36	5	33	4	2	31

Venus = 175						
22	47	16	41	10	35	4
5	23	48	17	42	11	29
30	6	24	49	18	36	12
13	31	7	25	43	19	37
38	14	32	1	26	44	20
21	39	8	33	2	27	45
46	15	40	9	34	3	28

Mercury = 260							
8	58	59	5	4	62	63	1
49	15	14	52	53	11	10	56
41	23	22	44	45	19	18	48
32	34	35	29	28	38	39	25
40	26	27	37	36	30	31	33
17	47	46	20	21	43	42	24
9	55	54	12	13	51	50	16
64	2	3	61	60	6	7	57

Moon = 369								
37	78	29	70	21	62	13	54	5
6	38	79	30	71	22	63	14	46
47	7	39	80	31	72	23	55	15
16	48	8	40	81	32	64	24	56
57	17	49	9	41	73	33	65	25
26	58	18	50	1	42	74	34	66
67	27	59	10	51	2	43	75	35
36	68	19	60	11	52	3	44	76
77	28	69	20	61	12	53	4	45

2.2. Symbolic Thinking and Sacral Numerology

A notable feature of the medieval world outlook is a symbolic perception of the environment when a symbolic function is attributed to everything – words, numbers, works of art, things, etc. According to Mažeikis, it is characteristic of human nature to look for a relationship between visible and invisible things (Mažeikis 1998: 14). Therefore, symbolic thinking was not alien to Antiquity: alongside the rational interpretation of the number as the main element of the beauty of the world when the Cosmos itself was the basis of everything, and the only truth, order and harmony, mythological meanings of numbers also functioned, which encoded the world of the antique Pantheon of Gods. The above-mentioned phenomenon of the harmony of spheres testifies to that – symbolic links of the planets with specific sounds, intervals or modes of music, numbers, the color, the geometric form or a vowel of the Greek alphabet.

Another example is the source of literature that has survived up to the present day, testifying to the mythological interpretation of numbers in ancient Greece.

This is the treatise *Theologumena Arithmeticae* (Arithmetical Theology, date unknown) attributed to the Neoplatonist philosopher Iamblichus. In this treatise teaching about numbers is treated not only from the position of mathematics, but also through the prism of myths presenting the corpus of the symbolism of the first ten numbers. According to the Russian philosopher Losev:

> [T]his is the only treatise that has survived up to the present day in which the theory of numbers is presented really extensively and in great detail on the basis of the Pythagorean school. (Losev 1988: 218)[62]

In the Middle Ages symbolic thinking began a new stage of perception of the surrounding world. The category of divinity that found itself in the centre of the transcendental truths of the medieval epoch replaced the picture of the rational and material image of the world and the principle of cosmocentrism formed in antique culture by the world of symbols and allegories, which was created by the Christian faith and religious patristic (Augustinian) symbolism. Attention was concentrated on one of the major witnesses of God – the Holy Writ, the basis of the philosophical thought of that epoch, the most significant source of investigations into, and explanations of, the medieval culture, the analysis of the texts that were compared to the process of perfection. Seeking to reveal the esoteric meanings encoded in the texts of the Old and New Testaments, they were given to even the simplest structural elements – monograms, colors, individual letters or their combinations, geometric shapes or numbers. By means of the latter, the life of the saints was ciphered, and each number could mean a specific act of sacrality. The Holy Writ was analyzed as the world of sacral numerical combinations and codes. All that turned into the rich symbolic language of numbers in the Middle Ages permeated with Christian sacrality.

Therefore, on the plane of interaction between music and mathematics, we encounter a qualitatively new semantic aspect. Musical examples from the early Christian period, such as the Gregorian chants, can be interpreted as a numerical symbolism of tones to comment on the Holy Bible. This complex of monodies was systematized as church canon in the 7th century by Catholic Pope Gregory I. It symbolically linked the sacred text with musical material. In the liturgical text, and in its structure, hidden meaning influenced the rules of the musical structure: the systematic repetition of individual words influenced the logic of melodic motifs; the musical material "reacted" to the appearance of particular holy words.

62 In the supplements to his books, Losev presents the translation of the treatise by Iamblichus and exhaustive commentaries on it (Losev 1988; Russian translation by V. V. Bibichin).

In the Christian numerological tradition there was a practice of applying meaning to numbers. The number three (the number of the Holy Trinity) had the most meaning. Also, the numbers 6 and 12 were significant. Six was the number of days it took to create the world (God created Earth in six days). Twelve was significant because of the number of apostles; the Christian Church symbol was associated with this number as well. For example, it has been noted that in the various levels of structures of the medieval mass – in the text as well as in the musical notations – the symbolism of numbers is encrypted with the essence of a Christian worldview. Especially the concept of the Holy Trinity is expressed through a variety of combinations of the sacred numbers. This is apparent in the symmetrical three-part structure of the musical text of *Kyrie* and *Agnus Dei*, which cannot be separated from its relationship with the liturgical text. The triple acclamation *Kyrie* (Greek *κύριος* – Lord) in the text is made up of three lines that are repeated three times:

Kýrie, eléison!
Christe, eléison!
Kýrie, eléison!

The *Agnus Dei* (Latin – Lamb of God) text has three lines:

Agnus Dei, qui tollis peccata mundi, miserere nobis.
Agnus Dei, qui tollis peccata mundi, miserere nobis.
Agnus Dei, qui tollis peccata mundi, dona nobis pacem.

In the opinion of Manfred Bukofzer, Baroque music was understood as the totality of mathematics, symbolism, and allegory (Bukofzer 1939/40: 21). The musical Baroque epoch was especially linked to the sacred Christian numerological tradition, and the compositional practice that accompanied it, as well as medieval mysticism. According to Rolf Dammann, in this epoch:

> [T]he belief that nature is replete with symbols and that man must learn to recognize and understand those symbols was of supreme importance. (Dammann 1967: 414)

The influence of the cosmological traditions of Antiquity were also very important to the Baroque worldview. Hints of mathematical procedures were interpreted as a metaphor for the return of Pythagorean time. According to Tobias Gravenhorst (see Gravenhorst 1995), in most Baroque musical compositions numbers accentuate regularity and the inner order of musical material. What becomes obvious are the typical planned formulas of symmetry, proportions, even the predictable structure of the compositions of that time.

Therefore, the use of symbols in the Baroque epoch differs from the Middle Ages that were characterized by the single-minded theologically influenced number notations. The Baroque worldview presented the multilateral understanding, which absorbed both a pagan and Christian worldview. For this reason, the Baroque creative practice combined meaning from Antiquity, the results of astrological observation, the signs of Pythagoras' harmony of the spheres, the proportions of perfect beauty, and the mystification of numbers from the Middle Ages, and Christian numerology based on Biblical meaning. The sum of these meanings optimally gave significance to the birth of the Godly Cosmos and the harmony of its micro- and macrostructural elements. An example would be the number of months in the astrological year based on the number of apostles. The cycle of the moon is made up of 28 days (4 weeks/phases x 7 days. To be exact, depending on whether we consider sidereal or synodic case, the moon month lasts 27–29 days). The numbers 4 and 7 in Christian numerology and the numerology of Antiquity have an especially important function: in the Antique tradition four phases of the Moon are mentioned; four seasons of the year; four directions of the world, and the main elements of the world; quadrivium; tetractys; the square; seven planets; seven days of the week or one phase of the Moon. In Christianity there are the four canonical Gospels; four points of the cross; God's name in Latin and Greek, and also Hebrew Tetragrammaton is written in four letters; according to John of Patmos (John the Revelator) in the Apocalypse Heavenly Jerusalem's plan is in the shape of a square; the seven gifts of the Holy Spirit; seven virtues; the seventh day, which is a day of rest after the creation of the world; the book with the seven seals; seven as the number of the faith and the Church; Maria's symbol, and so on.

The synthesis of the understanding of the symbolism of Antiquity with the Christian world is obvious in the musical art of the Baroque. The allegorical subtext of a musical composition cannot be separated from its numerical combinations or from the composition as a whole. For example, based on the historical argument, that in the Holy Mary Church in Lubeck, where Dietrich Buxtehude (1637–1707) was the organist, above the altar there was a large, ornate, astronomical clock, which provided the direct inspiration for the composer's Passacaglia for organ in D major, BuxWV 161. The Passacaglia structure is dominated by the astrological origins cipher 28 and the numbers that it is comprised of, like 4 and 7, the allegory of the changing phases of the Moon.[63] Having analyzed the Passacaglia, it was established that the composition's seven-tone theme is made up of four pitches *d – a – c-sharp – b-flat.* From a tonal perspective, the Passacaglia is divided into four sections; each section has seven variations. However, the number 28 is not only significant from the perspective of astronomical symbol-

63 More about the analysis of Passacaglia see the articles by Wurm (1984) and Kee (1984 & 1988).

ism. In charts of the Kabbalah it is the magical number for the triangle,[64] in the science of mathematics it is the second term of the sequence of perfect numbers.[65]

The number 6 illustrates an undeniable constructive and semantic influence in music. This number is interpreted as a symbol of nature (the hexagonal or six-sided structure of a crystal is perfection). The number 6 becomes even more significant when it is associated with the six days in which the world was created. This sign of the divine process of creation was one of the reasons why composers of the Baroque era often put six compositions into a cycle. For example, noteworthy are Kuhnau's cycle of six sonatas for keyboard on Biblical themes. Arcangello Corelli (1653–1713) divides his famous cycle of twelve (that is 6 x 2) *Concerti grossi* into two parts consisting of 6 pieces each. This cycle's opus number is 6. Corelli's cycle of 12 sonatas for violin and *continuo*, Op. 5, is divided into two parts of six sonatas each. This Italian composer is also the author of the six *Sonate a tre*, WoO 5–10, published in Amsterdam in 1714. Additionally, Georg Friedrich Händel's (1685–1759) cycle of twelve *Concerti grossi* is also noteworthy for having Opus 6.[66] Other composition cycles by Händel include six organ concertos, Op. 4 (HWV 289–294). They were created by the composer as interludes for oratorios, which were performed at the famous London Covent Garden. The "second collection" was called six organ concertos, Op. 7 (HWV 306–311); six marches (HWV 419); six trio sonatas, Op. 2 (HWV 386–391); six fugues for organ or keyboard (HWV 605–610).

In Bach's creative work, cycles of six compositions each are very common: the second collection for organ, *Schübler Chorales* (BWV 645–50), is made up of six preludes. The third collection is made up of 18 (6 x 3) chorales (*Die Achtzehn Grossen Orgelchoräle*, BWV 651–68). The fifth is made up of 24 (6 x 4) of Kirnberger's choral preludes (BWV 690–713). Also, there are six trio sonatas for organ (BWV 525–30) and six English and six French suites for keyboard (BWV 806–17), in addition to six Partitas for keyboard (BWV 825–30), and six small preludes for keyboard (BWV 933–8). In addition, there is a cycle of six sonatas and partitas for

---

64 28 is the sum of the first seven numbers 1 + 2 + 3 + 4 + 5 + 6 + 7.

65 The sequence of perfect numbers is Zarlino's number sequence 6, 28, 496, 8128, and 130816. In this sequence all the numbers are equal to the sum of its proper divisors. For example, 6 = 1 + 2 + 3 or 28 = 1 + 2 + 4 + 7 + 14.

66 John Walsh's publishing house published this cycle by Händel a few times while the composer was still living. In the second publication in 1741, Opus 6 was marked. It is believed that this was not accidental, because in 1715 this same publisher achieved success when it published the Corelli cycle opus 6.

violin (BWV 1001–6) and six sonatas for cello (BWV 1007–12), six Brandenburg Concertos (BWV 1046–51) and six Orchestral Suites (BWV 1066–71) and so on.

The numerical symbol 10 often appears in Renaissance and Baroque music as well (in addition to this number's equivalents, 100 and 1,000). This number connects the interpretation of the perfect number of Antiquity (the sum of the first four numbers 1 + 2 + 3 + 4) with the Christian concept of God's Ten Commandments. Among the many examples from music Girolamo Frescobaldi's (1583–1643) cycle *Cento Partite sopra Passacagli* (1639, Italian *cento* – one hundred) is noteworthy. Lodovico Viadana's (~1560–1627) vocal masterpiece *Cento concerti ecclesiastici* (1602–1609) is made up of 10 groups of concertos (*Concerti falsi bordoni passeggiati*) for each voice type (for soprano, for alto, for tenor, and for bass) and 20 double, 20 triple, and 20 quadruple concertos (*Concerti falsi bordoni*). In Heinrich Schütz's (1585–1672) collection *Geistliche Chormusik* the major tonality is chosen in 10 motets. According to Gravenhorst, the total of the entire piece is made up of one thousand measures (that is, 10 x 10 x 10; Gravenhorst 1995: 122).

2.3. Codes of the Numerical Alphabets in Music

The practice of composing music as a certain mathematical puzzle, which was widespread in Germany in the 16th–18th centuries, was also related to inserting codes of the numerical alphabet into a musical texture. Due to the wide possibilities it provided, this alphabet was regarded as one of the best measures to realize original musical ideas. Having chosen it as a tool of composing music, the possibility for the especially subjective intentions of a creator opened up. This technique of coding was applied with a much more individual and subjective aspect. For example, Mattheson, when speaking about his contemporary Johann Sebastian Bach, stated that the latter taught his pupil, Mizler, tricks of mathematical composing, including numerical encoding of the alphabet (Mattheson 1740: 231).

The numerical alphabet is a systematic model of analogizing letters and numbers; thereby a huge number of subjective signs can be incorporated into a musical texture – meaningful combinations of certain words, coded maxims, inserted autographs, dedications, glorifications, thanks, historiographic data, the names of the saints, etc. On the basis of the notes left by Bach's colleague, Picander,[67] Friedrich Smend, one of the most eminent investigators of Bach's creative legacy, stated that the composer used the numerical alphabet of a natural sequence from

67 Picander is a pseudonym of the German poet Christian Friedrich Henrici (1700–1764). He was the author of the librettos of many cantatas of Bach, as well as of the St Matthew Passion.

A – 1 to Z – 24. This is the most popular variant of a numerical alphabet used by musicologists when analyzing compositions of the Baroque period:[68]

A – 1	G – 7	N – 13	T – 19
B – 2	H – 8	O – 14	U, V – 20
C – 3	I, J – 9	P – 15	W – 21
D – 4	K – 10	Q – 16	X – 22
E – 5	L – 11	R – 17	Y – 23
F – 6	M – 12	S – 18	Z – 24

The numerical alphabet is presented in Tatlow 1991: 8, 133; Gravenhorst 1994: 50.[69]

It must be noted that the numerical alphabet did not have a particular theological function. It was applied as an independent mathematical code. Most of the time, the numerical alphabet was associated with an individual composer's intention to convey a specific name through tones. This phenomenon was mentioned as the innovation of Renaissance era composers. It was established that Johannes Ockeghem (1410/25–1497), Obrecht, Dufay, des Prez and Tinctoris used the numerical alphabet to encode their own names (Tatlow & Griffiths 2016). Dufay accomplished this in his ballade *Resvellies vous et faites chiere lye*, which was composed in 1423. When Vittoria di Lorenzo Colonna (the niece of Pope Martin V) married Carlo Malatesta their names were encoded within a musical composition using the numerical alphabet. This was established when a specific pattern was found within the structure of the musical composition. The ballade's length of 73 *brevis* matched with the gematria coding of surname Colonna (3 + 14 + 11 + 14 + 13 + 13 + 5 = 73). Structurally, the ballade is divided into three parts. The first part, together with the vocal text, is made up of 71 tones (the gematria code for the name Martin). The number of tones in the

68 By the way, the Greek triple system that was used less seldom also deserves to be mentioned. In this system numbers were chosen as equivalents to letters from 1 to 9, 10–90, and 100–900.

69 Tatlow presents this alphabet in the supplements alongside the other 32 variants of numerical alphabets. This is the first variant of a natural Latin order, which was recorded in their works by many theoreticians of the Baroque period, such as Rudolff (1525), Jacob (1565), Kuhnau (1700) (Tatlow 1991: 131–8). Gravenhorst mentioned the same alphabet too (Gravenhorst 1995: 50). He based himself on the 17th century source, the treatise *Delitiae mathematicae et physicae: der mathematischen und philosophischen Erquickstunden* (Delights of Mathematics and Physics, 1651) by Georg Philipp Harsdörffer (1607–1658).

second and third phrases for tenor and the contralto is 87 (the gematria code for the surname Malatesta).[70]

However, the Baroque era distinguishes itself for the large number of examples of the use of the numerical alphabet in musical compositions. At this time, there was a particularly great interest in using numerical code to personalize musical compositions. In addition to the personalized numerical ciphers, holy words, monograms, abbreviations, the numerological symbols of Christ and others were widely used. For example, the conscious use of the musical numerical alphabet can be witnessed in a work of Johann Christoph Faber (flourished in the early 18th c.), the nine-part *Neu-erfundene obligate Composition*. Within the structure of this composition there is a hidden code for the name Ludovicus, Duke Louis Rudolph of Brunswick-Lüneburg. The name is made up of nine letters, which matches the nine parts of the composition. At the beginning of each part, instead of the title, one letter from the name appears. That letter is "pronounced" as a musical tone in the trumpet solo. In the first part, in the place of a title, the letter "L" appears and the trumpet performs a 20-tone melody. In the second part, identified by the letter "U", there is a 200-tone melody, and so on. By comparing the letters with the number of tones, it is obvious that Faber was acquainted with the version of the Latin alphabet from A – 1 to I – 9, from K – 10 to S – 90, and from T – 100 to Z – 500.

Kuhnau's cyclical work on Biblical themes (six sonatas for keyboard, *Musicalische Vorstellung einiger biblischer Historien*, 1700) is an interesting example as well. In the cycle's annotation, the Baroque composer himself admits to using the numerical alphabet and explains how he encoded a certain name within his composition:

> However, should any be so curious as to want to know his name I shall convey it to him in an algebraic problem for his amusement and as *lusus ingenii* (nothing in addition that this entire work of mine, as my maiden muse on the first printed page clearly gives to understand, is nothing but such a *lusus*). But, first he must know that I have allotted every letter a number corresponding to its position in the alphabet: thus A equals 1, B equals 2, and so on. Further, I shall leave the reader in doubt as to whether I have used one or two letters too many or too few at the end, so that he should not immediately draw any conclusion from simple observation of the number of letters. Nevertheless, the name will appear clearly from the correct solution. This algebraic riddle, then, proceeds thus: the sum of the letters is a certain number.[71]

70 A comprehensive analysis of Dufay's *Resvellies vous* is presented in the article by Allan W. Atlas (see Atlas 1987).

71 In the quoted text Kuhnau highlighted the particular words himself. Translation by Michael Talbot, cited in Kuhnau 1700/1973: xi.

Wolfgang Reich, the editor of the facsimile publication of the Kuhnau sonatas, believes that the composer's rational tendencies for creation were dictated by the circumstances of his life. While studying at the Dresden Cross School (*Kreuzschule*) and at Leipzig University, he became not only a musician, but also a theologian, a judge, a rhetorician, a mathematician, and an expert of foreign languages. Therefore, having been so exceptionally gifted, Kuhnau possessed the virtuosity to manipulate and integrate various scholarly disciplines (Kuhnau 1700/1973: 33). Here is the riddle that Kuhnau presented in the annotation of his Biblical sonatas (Kuhnau 1700/1973: XI):

> … the sum of the letters is a certain number. The first letter would be a quarter of the total if it (the letter) were greater by 4. The next letter is too large by 8 to be one eighth of the whole. If 1 is added to the third letter it will be one third of the first. If one subtracts 4 from the sum of the remaining letters, it will have the same relation to the sum of the previous three letters that the three angles of a triangle have to two right angles. The fourth letter, however, is three times as great as its predecessor. And if 7 is added to the sum of these four letters, the fifth becomes the square root; in the same way as the sixth, if increased by 1, is the cube root of the fifth. If 2 is subtracted from the seventh letter and added to the eighth, each of these becomes one eighth of the sum total as mentioned above and still unknown.

An analysis of the riddle revealed that the hidden name was Stephani. This was the name of Agostino Stephani (or Steffani, 1654–1728).[72] When the chain of calculations was reconstructed, it became apparent that the composer used the traditional version of the Latin numerical alphabet, when A = 1, B = 2 and so on. When this alphabet was applied, the name Stephani was decoded. This also may explain the secret hidden in the composer's phrase "I have used one or two letters too many or too few." Possibly it means that Kuhnau chose not the commonly used Italian version of the name, Steffani, but the more rarely used Latin variant, Stephani (with two letters interchanged in the middle).

By applying the letter combination of the name STEPHANI, Kuhnau's riddle was solved as follows:

- the sum of the letters of the name equals 88:
  S = 18, T = 19, E = 5, P = 15, H = 8, A = 1, N=13, I = 9;

72 Agostino Stephani, an Italian clergyman and politician, a Baroque theoretician, and a composer of high standing, who lived in Munich, Hannover and Düsseldorf, was an important figure in German Baroque music at the time. In the musical panorama of the era his stature was likened to other great Italian musicians, such as Biagio Marini, Carlo Pallavicino and Antonio Sartorio.

- the first letter S is 18, therefore if 4 is added to 18 it will be 22, and 22 is ¼ of the entire sum (88);
- the second letter T is 19, therefore: 19–8 = 11, and 11 is ⅛ of the entire sum;
- the third letter E is 5, therefore: 5 + 1 = 6, and 6 is the ⅓ part of the first letter (S – 18);
- the sum of the five last letters P, H, A, N, and I is 46; by taking 4 away from 46, we get 42, the sum of the first three letters S, T, and E;
- the fourth letter P (15) is equal to the letter E (5) multiplied by three: 5 x 3 = 15;
- if 7 is added to the sum of the first four letters S, T, E, and P – 57, we get 64, and 64 is the letter H (8) squared, because 8 x 8 (or 8^2) is 64;
- if 1 is added to the sixth letter A (1) we get 2, and 2 is the cube root of letter H (8), which comes before letter A, therefore: 2 x 2 x 2 or $2^3 = 8$;
- if 2 is substracted from the seventh letter N (13), or added to the final letter, I (9), we achieve the same result – 11, and 11 is the ⅛ part of the sum of all the letters in the name.

Analysis of Biblical sonatas revealed that the number 88, and every letter's numerical equivalent, influences the construction of the entire cycle. For example, the first sonata is made up of eight parts – the name Stephani has as many letters, and the numerical equivalent of the letter H; moreover, this sonata is made up of 22 structural sections – that is ¼ of 88. The sum of the parts of sonata pairs No. 1 & 2 and No. 5 & 6 matches 11 each pair (8 and 3 as well as 6 and 5), which is often shortened by Kuhnau to ⅛ (part of the number 88). The sixth sonata is made up of 11 sections – the same shortening by ⅛. The sum of the sections of the first three sonatas are 18 (8, 3 and 7) which is the match for S. The third sonata's sections is the number 15, which matches the letter P. The number of sections of the last sonata (No. 6) is the number 5, which is the equivalent for the letter E.

Further I analyzed the numerological connections with the heroes of the Holy Bible, who were mentioned in the sonatas. For example, it was established that the numerical code for Stephani, 88, matches for the names of the two main characters of the second biblical sonata *Der von David vermittelst der Music curirte Saul* (Saul's Cure through Music by David) – the Israeli shepherd boy, David, and the name of King Saul. The sum of the numerical equivalents of the letters of the names is 88 (David 38 + Saul 50). The number of parts of this sonata, three, most likely is an allegory for the three gifts that are mentioned in the 16th chapter of the Book of Samuel:

> So, Jesse took a donkey loaded with bread, a skin of wine, and a young goat, and sent them with his son David to Saul. (I Sam 16: 20)

This hypothesis allows one to reveal an analogical coincidence in the first sonata, *Der Streit zwischen David und Goliath* (The Combat between David and Goliath): why did Kuhnau chose this Biblical story for the theme of his composition? It could be because the number 5 (the equivalent for the letter E) is mentioned in the sacred text about the fight between David and Goliath:

> Then he took his staff in his hand, chose five smooth stones from the stream, put them in the pouch of his shepherd's bag, and, with his sling in his hand, approached the Philistine. (I Sam 17: 40)

Moreover, the number 5 in this sonata organizes the rhythmic image of the fight between David and Goliath: the music is based on two characteristic five-tone rhythmic formulas. In the score the second formula appears 10 (5 x 2) times. Another coincidence is that in the final three measures C major is established by a harmonic material of 69 tones, which is the numerical equivalent of the number of letters of the name *Goliath*.

The use of the number seven in the third sonata, *Jacobs Heyrath* (Jacob's Wedding), can be explained according to the same principle: it is a link to the sacred text. The seven parts of the sonata are possibly an allegory for the numerical symbol from Genesis. It represents the agreement between Jacob and Laban:

> So Jacob served seven years to get Rachel, but they seemed like only a few days to him because of his love for her. (Genesis 29: 20)

The three-part structure of the fourth sonata, *Der todtkrancke und wieder gesunde Hiskias* (Hezekiah's Mortal Illness and Recovery), can be connected with the excerpt from the second book of Kings, where it is written that King Ezekiel prayed fervently, and on the third day was healed. Additionally, in the story of Ezekiel's illness, the fourth letter, P, of the name Stephani, with its numerical equivalent 15, takes on special meaning:

> I will add fifteen years to your life. And I will deliver you and this city [...]. (2 Kings 20: 6).

The tonal and alphabetical manifestation of Bach's name and surname pays a lot of attention of researchers as well. This surname stands out for the fatalistic sound of the intersected minor seconds, which semantically create the graphic of the cross, and with the possibilities of gematria calculations of the name and surname, which are dictated by the Latin variant of the numerical alphabet. An anthologized example in music of the encoding of the composer's name can be found in the theme of the Fugue in C major (WTC Vol. 1, BWV 846). Because of the structure of its 14 tones, it is often interpreted as Bach's musical autograph (see Figure 14). Other numerical cases in Bach's creation are mentioned too. For example, the length of *Credo* part

(Latin – profession of faith) from the mass in B minor, BWV 232, is not accidental either. The numerical code of the word "Credo" is 43: C + R + E + D + O = 3 + 17 + 5 + 4 + 14 = 43. The sum of the number of measures of the part, 129, allegorically symbolizes three times chanted "Credo, Credo, Credo" (because 129 = 3 x 43); the same word, "Credo" in this part is repeated by the composer a total of 43 times.

The use of the numerical alphabet to encode personal messages was also used in Classical period music. For example, Martha Frese developed a hypothesis about Mozart's last opera, *The Magic Flute* (1791) stating that characters from the Josephinian Vienna time period were immortalized in the opera. Political functionaries of Vienna were identified using Masonic interpretation and the numerical alphabet in the structure of the opera. The musical material combinations encrypted the Masonic magistrate Ignaz von Born, Emperor Joseph II, and the Austrian Empress, the Queen of Hungary and Bohemia, Maria Theresa (more see Frese 1998).

Figure 14. Various numerological combinations for the name Bach; a manuscript fragment with the 14-tone subject of Fugue in C major, BWV 846 (the information was systematized by the author of this book – R. P.*)*

Bach	**14** (digital root 5)[73]
J. S. Bach	**41** (root 5)
J. Bach	**23** (root 5)
S. Bach	**32** (root 5)
Johann Sebastian Bach	**158** (root 5)
J. S. B.	**29** (same for the monogram SDG)
Johann Bach	**72**
Sebastian Bach	**103**

73 The digital root (or repeated digital sum), or the "modulo 9" rule. The digital root is established by adding digits of an integer repeatedly, until a single digit is arrived at. For example, the digital root of the number 444 is 3, and this is achieved by: 4 + 4 + 4 = 12 → 1 + 2 = 3. On the other hand, it is an elementary mathematical fact that the digital root of an integer *n* is exactly its non-negative reminder if reduced by modulo 9. In Bach's case the number 14 has a root 5, which we get by adding up its members 1 and 4: 1 + 4 = 5.

Figure 15. Numerical equivalents of sacral words, names and monograms and for the versions of Christ name (the information was systematized by the author of this book – R. P.*)*

Α–Ω (α–ω) *Alpha* and *Omega* (the first and final letters in Greek alphabet) – the beginning and the end, God Father and Son	**1 & 24**	IGNIS	**56**
		MARIA	**40**
		MORS	**61**
		MRA (shortcut to *Maria*)	**30**
		PATER	**57**
AERIS	**50**	SANCTUS	**92**
AQUA	**38**	SDG (*Soli Deo Gloria*)	**29**
CREDO	**43**	SPIRITUS	**125**
CRUCIFIXUS	**127**	SS (*Sanctus Spiritus*)	**36**
CRUX	**62**	T (Greek *Theos*)	**19**
DEO	**23**	TERRA	**59**
DEUS	**47**	TRINITAS	**105**
FILIUS	**73**	TRI-UNITAS	**125**
GEIST	**58**	UNITAS	**80**
GLORIA	**59**	VIRGO	**67**
GOTT	**59**	VITA	**49**

IHC (*Ihcuc*)	**20**
JJ (*Jesu Juva*)	**18**
J.CHR	**37**
JESU	**52**
JESUS	**70**
CHRIST	**74**
CHRISTE	**79**
CHRISTUS	**112**
JESU CHRISTE	**131**
JESUS CHRIST	**144**

XP – **37**	The first two letter of Christ's name in Greek language *ΧΡΙΣΤΟΣ* (*Χριστός*). Their combination is a symbol of cross. The exchange of letters to *PX* derives the shortcut for *Pax* (Latin – peace).
ICXC – **37**	The ancient monogram: *IC* (12, the first and final letters for *Ihcuc*)and *XC* (25, the first and final letter for *Xpictoc*)
NIKA – **33**	The Greek version of Lord
IHS / JHS – **35**	A monogram for: a) Greek *ΙΗΣΥΣ*, b) Latin *In hoc signo*, and c) *Jesus Hominum Salvator*
INRI – **48**	*Jesus Nazarenus Rex Judaeorum*

The examples I have discussed show how the manifestation of the numerical alphabet in a musical composition is a constructive musical and verbal dialogue in which numbers serve as mediators. In the musical practice of the 19th century another form of this dialogue took hold – writing hidden messages within a musical composition using cryptography,[74] while in the 20th century this technique was enriched and supplemented by the constructive use of the Morse Alphabet.[75]

74 The theoretical basis of composing cryptographic music and the first actual examples of cryptographic musical systems are dated from the beginning of the 12th century.

75 The applications of cryptography and Morse code to music are presented in the section "Musical Cryptography as a Common Denominator of the Sound and the Word" in this book, pp. 89–92.

3. Constructive Aspects of the Interaction between Music and other Arts

The conception of mathematically perceived beauty, its expression in terms of numerical manipulations, as well as the laws of symmetry and proportions as formulas of beauty, could have been a connecting chain revealing the links between music and other arts, such as architecture, literature, and painting. In these fields mathematics exists as a common denominator and is an element of the constructive tie between the contrasted and compared arts. The same proportions of numbers in different epochs were regarded as the condition for perfect art in music, architecture, fine arts, and poetry.

3.1. "Frozen Music": Dialogues between Music and Architecture

The issue of the affinity between music and architecture has been raised by many thinkers from different historical epochs. In the early aesthetics of Antiquity architectural buildings that corresponded to numerical relations of musical intervals were likened to an euphonious sound of a musical chord. For example, the antique theory of the impact (or the influence) of music harmony on the structure of the world is illustrated by the ancient Greek myth about the Amphion,[76] son of Zeus and Antiope, testifying to the relationship between music and architecture. When Amphion played the lyre that Hermes gave him, the magic stones and beams gently glided into place by themselves, and the walls of the city of Thebes were built.

The representative of Hellenism, Vitruvius, thought that a good architect should understand music:

> Music, also, the architect should understand so that he may have knowledge of the canonical and mathematical theory, and besides be able to tune ballistae, catapultae, and scorpiones to the proper key.[77]

---

76 The Greek traveller and geographer Pausanias (Παυσανίας, 2nd c. A.D.) wrote in his *Description of Greece* that Amphion was Hermes' pupil, and the first player of the lyre. He improved the lyre by adding three more strings to the former four, and his singing and playing moved stones and made animals follow him (according to Pausanias: *Description of Greece*, Book IX, Chapter 5).

77 Cited in Vitruvius: *The Ten Books on Architecture*, Morris Hicky Morgan (transl.), Cambridge: Harvard University Press, 1914, Book I, I.8, p. 8.

Somewhat later, the philosopher and theologian St. Augustine (354–430) called architecture and music sisters – "children of the number" (Warner Marien & Fleming 2005: 210). The representative of Italian Renaissance architecture, Leon Battista Alberti[78] (1404–1472), in his treatise *De re aedificatoria* (On the Art of Building, 1443–52) described the relationship of proportions used in architectural art with the proportions of musical intervals established by the followers of Pythagorism. In music, proportions of numbers manifest themselves as intervals, and in architecture as planes (of squares, forums) or measurements of buildings (the Palace of the Senate, halls and others). Alberti presented examples illustrating that squares or streets should be designed observing the following proportions 2 : 3, 3 : 4, 1 : 2, 1 : 3, 1: 4, 8 : 9 (more see: Saleh Pascha 2004, Strohmayer 2001). We remember that these numerical formulas match with the proportions of musical intervals set by Pythagoras: a fifth 2 : 3, a fourth 3 : 4, an eighth 1 : 2, and a whole tone 8 : 9.

Another Italian Renaissance architect, Andrea Palladio (1508–1580), took a similar position. In his *I quattro libri dell'architettura* (The Four Books of Architecture, published in 1570) he described the main numerical proportions that architects should follow – they are equivalents of the main musical intervals (Saleh Pascha 2004: 75).

The works of German thinkers of the 19th century reveal an especially intense interest that philosophers took in the issues of the interaction between music and architecture.[79] During this period, the metaphor "frozen music" emerged.[80] Its authorship is attributed to the representative of German idealism Friedrich Wilhelm Joseph Schelling (1775–1854). In a comprehensive study Saleh Pascha (Saleh Pascha 2004) noted that, according to Schelling's

78 By the way, the ideas about a general theory of beauty (the so-called *pulchritudo*) based on numbers, proportions and order (*numerum finitionem collocationemque*), that is, an elegant combination (*concinnitas*) of different elements corresponding to musical proportions, belongs to this Italian architect of the 15th century.

79 In the Romantic period some analytical studies were published, such as John Moore Capes' article "Music and Architecture" (in *Fortnightly Review*, 1867), on analogies between gothic cathedrals and musical counterpoint compositions, or W. Schultz's *Die Harmonie in der Baukunst* (Harmony of the Architecture, 1891) on the interaction between proportions and musical intervals in architecture.

80 The German concept *Gefrorene Musik*, a word-for-word translation into English is *Frozen Music*.

Philosophie der Kunst (Philosophy of Art, 1802–3), music and architecture were related allegorical arts, moreover, architecture is music in space, a frozen music.[81] Another representative of that time, Hegel (1770–1831), specified that the common feature, the essence of both kinds of art – architecture and music – was their numerical substantiation (German *Zahlenge-setzlichkeit*) (Saleh Pascha 2004: 128), although, according to Hegel, these two arts represent entirely opposing categories. Goethe (1749–1832) left the following aphorisms of architecture in his diary in 1827 and 1828: "silenced art of sounds" (German *verstummte Tonkunst*), "mute music" (German *stumme Musik*), "frozen music" (German *erstarrte Musik*). Madame de Staël (1766–1817), inspired by her trip to India, wrote in her novel *Corinne* (1807) the following: "Looking at this monument is as though hearing continuous and immovable music."[82] And vice versa, the term architecture was borrowed and applied to the world of music as well, for example "melted/slackened architecture" (German *aufgetautete Baukunst*, Friedrich Theodor Vischer's maxim in the 1846–1857 treatise) or "fluid/smelted architecture" (German *flüssige Architektur*, August Wilhelm Ambos' idea, 1855).

Yet, the interpretation of the interaction of these art forms in works of the Romantic era were not unilateral. For example, in the four Romantic era depictions of the hierarchy of art (German *System der Künste*) the positions of architecture and music were regarded differently. Schelling and Vischer defined the rules of these arts as the foundation for the other arts. Meanwhile, Hegel and Karl Wilhelm Ferdinand Solger (1780–1819) emphasized a contrast between architecture and music as opposing art forms of symbolism and Romanticism (see Figure 16).

---

81 Original quotation in German, published in Friedrich Wilhelm Joseph Schelling's *Philosophie der Kunst* (in: *Sämtliche Werke*, Abt. 1, Bd. 5, Stuttgart, 1859, p. 0252: 576):

> Die Architektur, als die Musik der Plastik, folgt also nothwendig arithmetischen Verhältnissen, da sie aber die Musik im Raume, gleichsam die erstarrte Musik ist, so sind diese Verhältnisse zugleich geometrische Verhältnisse.

Besides, Saleh Pascha traces the origin of the concept itself in the myths of ancient India, China or Egypt, which speak about the creation of the world as silencing (= putting a stop to) the initial sound and its conversion into matter (Saleh Pascha 2004: 5).

82 Original in French (in Madame de Staël's *Corinne*, Book 4, Chapter 3):

> La vue d'un tel monument est comme une musique continuelle et fixée.

Figure 16. Different versions of System of Arts according to Schelling, Vischer, Solger and Hegel (the diagrams reproduced from Saleh Pascha 2004: 100, 114, 128, and 141)

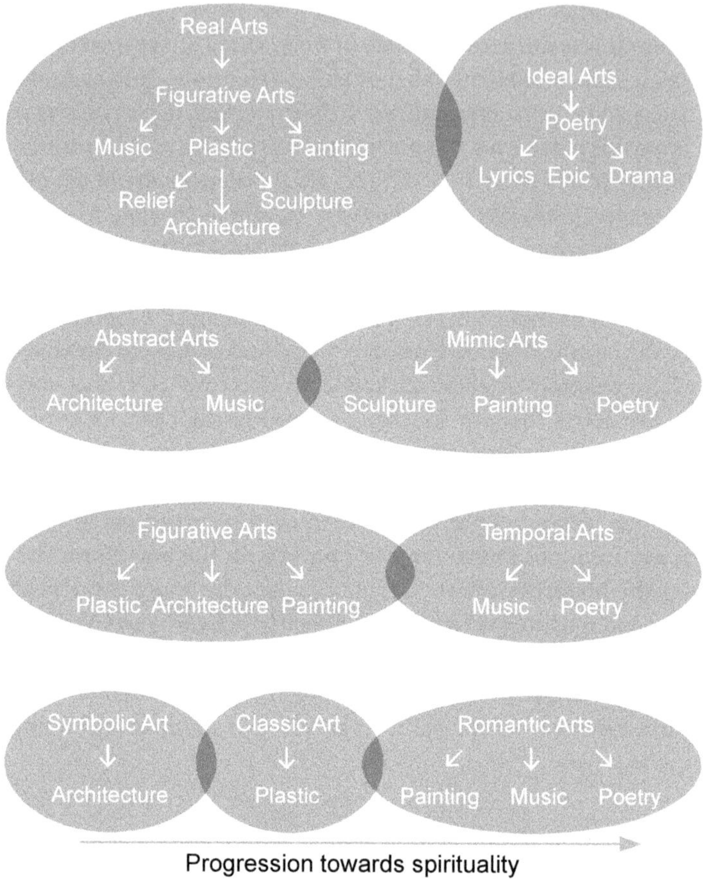

Thousands of years ago, the idea of a common genesis for music and architecture was raised, and that idea of the interaction of these two art forms still persisted in the 20th century. This is proven by a few important works of this time period: Rudolf Wittkower's *Architectural Principles in the Age of Humanismus* (1949), which is one of the most important works about the influence of musical thinking and musical theory on well-known Renaissance architects (for example, Alberti, Palladio); Hans Kayser's work *Harmonikale Proportionen in der Baukunst* (1958), which analyzes the parallels between architecture and music; Paulo von Naredi-Rainer's *Architektur und Harmonie* (1982); Michael Bright's *Cities built to*

Music (1984); Stefan Fellner's *Numerus sonorus. Musikalische Proportionen und Zahlenästhetik in der Architektur…* (1993); Peter Bienz's *Le Corbusier und die Musik* (1998) or Fritz Neumeyer's *Der Klang der Steine* (2001).

The discussion over the interaction between architecture and music over the course of epochs was always in counterpoint with the fact that all of it could be materialized in architectural structures or in musical compositions, which sought a visual expression. For example, while studying the artifacts of the Temple of Apollo in the antique city of Didyma in southern Turkey, Jens Birnbaum based the building's measurements on musical theory and established that the proportions of the ancient temple were equivalent to the numerical expression of musical intervals as described by Pythagoras (more see Birnbaum 2006). Pozzi Escot argues that the numerical relationships of musical intervals are obvious in the measurements of gothic cathedrals: the numerical expression of the octave is 1 : 2, and it matches with the relationship of the measurements of the length and width of the church's foundation, a fifth interval is expressed as 2 : 3, which matches the relationship between the church's general length and the transept (or nave), the fourth interval's formula 3 : 4 sets the measurements of the church's choir, while the major third interval 4 : 5 "rings" when measuring the distance between the church nave and the side altars (which are considered a whole). Escot saw an analogy between the architecture of that time and the compositions of the medieval composer Hildegard von Bingen (1098–1179). Having analyzed the four antiphons' structure, by applying the method of graphic analysis she created the scale plan for the psalm antiphon *O pulcre facies* and compared the graphic expression with the construction typical with facades of gothic churches (Escot 1999: 9–16).

Renaissance architectural monuments are often analyzed as "sounding images of music." An example would be the building of the Cathedral of St. Francis in Rimini (which is still called Tempio Malatestiano). This 13th century Gothic cathedral was partially reconstructed by the architect Alberti in c. 1450. Barbara Barthelmes argues (1985) that the measurements of the cathedral reflect the numerical relationships of musical intervals. The same architect is attributed with the 1451 construction of the Rucellai Palace in Florence, which perfectly illustrates Alberti's theory about the relationship between architectural measurements and musical intervals.

An example of how an architectural structure inspired Renaissance music is the English composer Leonel Power's (1370/1385–1445) mass *Alma Redemptoris mater*. It was noticed that this composition's *cantus firmus* and the relationship with other musical voices, is expressed in the numerical sequence 48 : 12 : 48 : 24

(a reduction is 4 : 1 : 4 : 2). This sequence matched with the measurements of the upper facade of Alberti's design of the Basilica of Santa Maria Novella in Florence (Tatlow & Griffiths 2016).

An especially unique connection between architecture and music based on a numerical foundation is illustrated in the motet of Renaissance composer Dufay *Nuper Rosarum Flores* (1436). This festive composition was dedicated in honor of the March 25, 1436 blessing of the Florentine Cathedral of Santa Maria del Fiore. According to researchers (for example see: Warren 1973, Ryschawy & Stoll 1988, Wright 1994), the form of musical material and the rhythmic proportions repeat the cathedral's architectural measurements. The isorhythmic motet is made up of four sections with a gradual mensural change. This mensural change is expressed in the numerical sequence 6 : 4 : 2 : 3; the first section is written as *tempus perfectum* (the time signature is 6/2), the second – *tempus imperfectum* (time signature 4/2), the third – *tempus imperfectum diminutum* (time signature 4/4, or 2/2) and the fourth – *tempus perfectum diminutum* (time signature 6/4, or 3/2). This number sequence matches the proportions of the cathedral cupola designed by architect Filippo Brunelleschi (1377–1446), i.e. these numbers are manifested in the measurements of the cathedral's nave, crossing, apse and the height of the dome.

We can find examples of architectural reflection in musical compositions from later epochs too. In the opinion of Wolfgang Stechow, Modest Mussorgsky's (1839–1881) "The Great Gate of Kiev" (from piano cycle *Pictures at an Exhibition*, 1874) is not only a direct transcript of a painting by Viktor Hartmann, but more likely a sound allusion to an architectural ensemble. And the sound of Debussy's prelude *La cathédrale engloutie* (The Submerged Cathedral, 1910) affects one like a strong architectural symbol or like Monet's painting of the Rouen Cathedral (Stechow 1953: 324).

An especially original instance of an architectural "sound" can be found in a 20th century musical composition's visuality. For example, Martin Gardner shows that the Brazilian composer Heitor Villa-Lobos' (1887–1959) composition for piano *New York Skyline Melody* (W 407, 1939) was inspired by a graphic image – the shape of New York City panorama. This image was transposed into a graph paper using the so-called millimetrization method, which allows for the appropriate tonal pitches. An analogical method of analysis can be applied to Sergei Prokofiev's (1891–1953) music score for Sergei Eisenstein's film *Alexander Nevsky* (1938): the silhouettes of people and landscapes from the upcoming movie were used by the composer to place tonal pitches on the staff (Gardner 1974: 134–6 & 1992: 22). According to Gregg Wager, through the workroom window of Karlheinz Stockhausen (1928–2007) in Paspels, Switzerland, he could see the contour of a mountain range, which became the tonal pitch archetype for his composition *Gruppen* for three orchestras (1957), while "envelopes of rhythmic blocks are exact lines of mountains" (Wager 1998: 84). The visual prototype also determined Larry Austin's (born 1930) fractal canons *Canadian Coastlines* for performers and for a computerized ensemble (1981). A computer generated tonal pitches, rhythm, tempo, timbre, and dynamics, based on the algorithms of the coordinates of the Canadian coastline (Dodge 1988: 10).

In contemporary music particular architects and their work are cited as sources of creative inspiration. Ideas for filling non-traditional works and spaces by one of the most interesting of contemporary architects, Daniel Libeskind,[83] inspired the composer Simon Bainbridge (born 1952) to write his composition *Music Space Reflection* for an instrumental ensemble and for electronics (2006). This piece for the composer is a tonal reflection of an experimental visual structure. The result of the interaction of music and architecture in a 20th century retrospective can be seen in the architect Le Corbusier building – the Philips pavilion. Edgard Varèse (1883–1965) composed his *Poème électronique* for tape (1958) as this building's 11-channel "recording." That same pavilion's hyperbolic construction inspired Xenakis who worked with the project to create the structural decisions for orchestral *Metastasis* (1953–4) – the image of the Philips pavilion was "transferred" to the score, e.g. for the strings *glissando* graphic.

3.2. *Ut Pictura Musica*: The Interaction between Music and Art

The numeric interaction between music and fine arts was already discussed in the treatises of the Renaissance period. These treatises mention the conception of *ut pictura musica* that revealed parallels between music and painting, and functioned in parallel with the definition of art and poetry *ut pictura poesis*.[84] For example, in the conviction of the Italian painter and theoretician Giano Paolo Lomazzo

---

83 Daniel Libeskind (born 1946 in Poland, resides in the United States) is an architect, artist, and set designer. Before he became interested in architecture, he studied music. As an architect, he designed unique buildings, such as the Jewish Museum in Berlin, the Imperial War Museum North in Manchester, and the Royal Ontario Museum in Toronto.

84 Italian *ut pictura musica* – music is painting, and painting is music. The maxim *ut pictura poesis* is a clipping from *poema pictura loquens, pictura poema silens* (Latin – poetry is speaking pictures, painting is mute poetry). In the Renaissance *ut pictura poesis* and *ut pictura musica* were main conceptions considering interrelations between the arts, alongside them *ut pictura biographia* (analogies between the art of biographies and portrait painting) is to be mentioned.

(1538–1592), secrets of music perceived by artists determined the mastery of proportions peculiar to the creative work of such masters as Leonardo da Vinci (1452–1519), Michelangelo (1475–1564) or his teacher, the famous painter of the Renaissance, Gaudenzio Ferrari (1475/80–1546). This was due to the belief that "the painter cannot be perfect"[85] unless he understands the art of music. The Florentine painter Giovanni Balducci (c. 1560–c. 1600) regarded the proportions of numbers as the basis of music and painting and called these two arts sisters whose mother was mathematics. This was expressed as follows:

> [J]ust as the composer ensures that the low [bass] and high [soprano] voices are in proper proportion to other voices in the middle [alto and tenor], so too must the painter take care that he makes well-proportioned figures, that he doesn't join too big a head or pair of legs, that the other parts are appropriate to each other. (Korrick 2003: 200)

Numbers that functioned as common denominators were also an important part of the discussion of the analogy between musical tones and colors. The division of colors into seven groups in Ancient Greece influenced the theory of the harmony of the spheres and the link between colors and the seven musical tones, as well as the seven orbiting heavenly bodies that were known at that time. The theory of color harmony based on musical sounds formed within the context of the harmony of the spheres, or the so-called *spectrum-octave*, was significant up until the 16th and 17th centuries. The cosmological interpretation of colors was continued by polymath Hieronymus Cardanus (Girolamo Cardano, 1501–1576). In his encyclopedic study *De subtilitate* (On Subtlety, 1550) he connected the seven colors with the seven senses of taste and with the heavenly bodies. At the same time, in the fundamental treatise about harmony Zarlino argued that musical intervals and colors interact and that the intervals of perfect prime and the octave correspond to the white and black colors. The intermediate ranges (a fifth, a fourth, and a third) correspond to the green, red, and blue colors.[86] Marin Cureau de la Chambre (1594–1669) suggested taking the system of the harmony of music and the proportions of musical intervals and carrying them over into the relationships between colors (treatise *Nouvelles observations et conjectures sur la nature de l'iris*, 1640). Isaac Newton (1642–1727) wrote about the analogies between colors and musical tones and heavenly bodies (treatise *Hypothesis of*

---

85 Italian quotation "Non può essere perfetto il pittore" is from Lomazzo's treatise *Trattato dell'arte della pittura, scoltura et architettura* (Treatise on the Art of Painting, Sculpture, and Architecture, 1584; also see Korrick 2003: 201, 206).

86 Zarlino writes about this in the eighth section of the third part of his treatise *Le istitutioni harmoniche* (Zarlino 1562: 155).

Light, 1675). Kircher created a chart of the analogies between colors and musical tones, which he published in his treatise *Ars magna lucis et umbrae* (The Great Art of Light and Shadow, 1646). Also mentioned are Kircher's experiments with creating atmospheric flashes of a variety of colors as music was playing. One of the most active protagonists of this idea was the scholar and Jesuit Louis Bertrand Castel (1688–1757). He attempted to construct a light/color harpsichord (*clavecin oculaire*); this idea was brought to life in 1893 by the British artist Alexander W. Rimington, when he created a color organ.

The early 20th century synesthesia experiments of Alexander Scriabin (1872–1915) were based on Newton's color theory. In the score of his work *Prometheus*, Op. 60 (1910), there is a part for light (*Luce*). It is also widely known that Olivier Messiaen (1908–1992) made sound consonances synonymous with color: an A major sixth chord sounded to the composer like the color of the light blue Mediterranean Sea, or the color of a blue sky (Griffiths/Messiaen 2001: 496). The composer's score for *Couleurs de la cité céleste* for a low brass and percussion orchestra (Colors of the Celestial City, 1963) connected the musical chords with certain colors, which were chosen from the texts of the Book of the Revelation (Schouest 2000: 6). Constructive links between music and color are illustrated in Sofia Gubaidulina's (born 1931) composition *Alleluja* (1990). According to the composer, the law of physics by which color is absorbed upwards is expressed through numerical proportions and determined the structural dimensions of the composition: parameter of length, relationships of the structural parts (Cenova 2000: 25–7).

The expression of tonal colors was significant to Stockenhausen's seven opera cycle *Licht: Die Sieben Tage der Woche* (1980–2003). Each opera was created to adequately reflect each day of the week. In this cycle, the composer identified each day symbolically with an appropriate color: Thursday was blue, Saturday was black, Monday was light green, Tuesday was red, Friday was orange, Wednesday was light yellow, and Sunday was gold. According to Wager, surviving among the notes to the opera there is an entire system of the various elements; for example, each day of the week represented not only a certain color, but also a mineral, a gemstone, a plant, an animal, an ethnic nation, a planet, and a star (Wager 1998: 112–3).

Examples of a close tie between music and art, without a doubt, are identified with the creative work of prominent Lithuanian composer and artist Mikalojus Konstantinas Čiurlionis (1875–1911). His paintings were often created according to the principles of music composing,[87] and in his musical compositions one sees what

87 For example, Dorothee Eberlein researched the structural connections in Čiurlionis' two cycles on the theme of the sea in both his music and in his painting (see Eberlein

may be referred to as a graphic iconography. These parallels between the melodic line of the composer's unfinished and unpublished variations *Easacas* for piano (1906)[88] and the graphical curves in the trilogy of paintings completed that same year, *Kibirkštys* (Sparks, 1906), were studied by musicologists Vytautas Landsbergis and Darius Kučinskas (Landsbergis 1976: 210–6; Kučinskas 2004: 112; see Figure 17).

Figure 17. Above: Čiurlionis' paintings from the cycle Sparks *(1906) and their resemblance to the musical scales; below: fragment of variations* Easacas *for piano (1906) from Čiurlionis' manuscript (stored in the archive of the M. K. Čiurlionis National Museum of Art, Čm–21, page 211)*

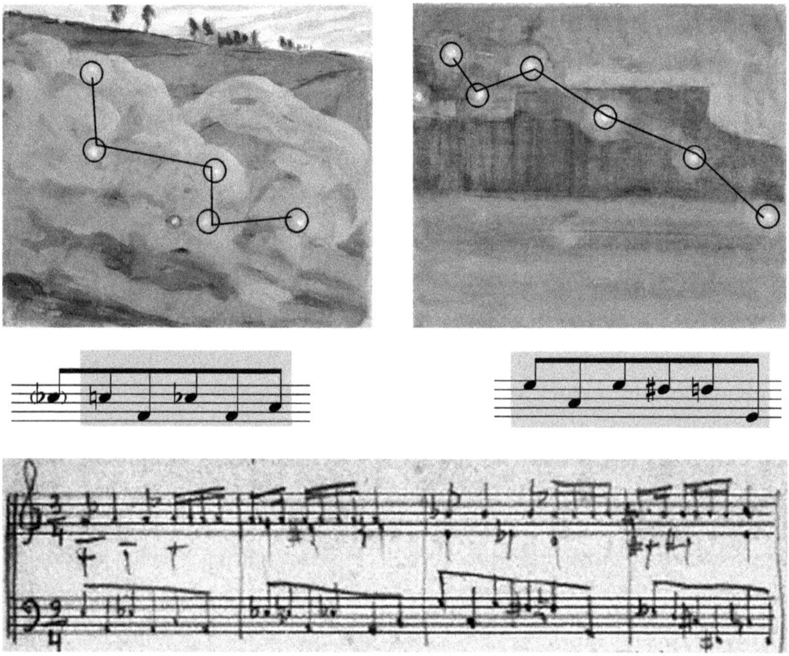

1994). Based on the affirmation of the Russian musicologist Vladimir Fedotov, the logic of the composer's paintings clearly repeats the logic of the forms of musical composition (see Fedotov 1989).

88 A theme for variation is composed of the tones mentioned in the title *Easacas* (*e – a-flat – a – c – a-flat*). According to Kučinskas, the origin of the alphabetical title is not completely known, but it is surmised that the code spells out Čiurlionis' name *MIkołAj konStAnty CzurlAniS*: MI as *e*, AS as *a-flat*, A as *a*, C as *c*, and AS as *a-flat* (Kučinskas 2004: 108). For more on this topic, see the section "Musical Cryptography as a Common Denominator of the Sound and the Word" in this book, pp. 79–92.

3.3. Musical Cryptography as a Common Denominator of the Sound and the Word

The sources of the practice of the cryptographic[89] composing of music as conceptual considerations of a constructive dialogue between musical and verbal arts are linked to the poet-musicians, Amphion and Orpheus, mentioned in antique sources, and the Biblical kings David and Solomon. It was in the considerations of the Renaissance writers that attention was turned to the antique examples by discussing commonalities of musical and verbal arts. Thomas Campion (1567–1620) called poets *artem qui tractant musicam* (Latin – those who toil in the art of music) and stated that the world is based on the symmetry and proportions arising from music and turning into poetry – there symmetry and proportions determine the metric and prosodic parameters of verses.[90] The English writer George Puttenham (1529–1590) in his treatise *The Arte of English Poesie* (1589) indicated that the

89 Cryptography (also, cryptology, Greek *κρυπτός* – hidden, secret + Greek *γράφειν* – writing, script) is the coding or ciphering of a message or information. Two ways of secret writing are distinguished during which the following is created: a random combination of letters/numbers, which has a certain meaning giving another meaning to an ordinary word/phrase; 2) the systematic changing of the placing of letters, replacing letters with other letters and symbols.
Mentioning the practice of secret messages goes back to the times of ancient Greece. For example, the system of ciphering messages, the Polybius square, is described in the work *The Histories*, consisting of 40 volumes by the historian, writer and politician Polybius (Πολύβιος, c. 203–c. 120 B.C.). Every letter is matched to a combination of two numbers that are arranged in the cells of the square (Polybius: *The Histories*, Book X, Chapter 45). For example, the square of the Latin alphabet and numbers is made on the principle of the Polybius Square:

	1	**2**	**3**	**4**	**5**
1	A	B	C	D	E
2	F	G	H	I/J	K
3	L	M	N	O	P
4	Q	R	S	T	U
5	V	W	X	Y	Z

90 Original quote "The world is made by Simmetry and proportion, and is in that respect compared to Musick, and Musick to Poetry: for Terence saith speaking of Poets, artem qui tractant musicam, confounding musick and Poesy together" was written in Thomas Campion's *Observations in the Art of English Poesie* (Chapter 1 "The First Chapter, Intreating of Numbers in Generall", 1602, page not indicated).

major feature of poetry is harmonious speaking and writing achieved by means of proportions transmitted from the art of music:

> Of all which we leaue to speake, returning to our poeticall proportion, which holdeth of the Musical, because as we sayd before Poesie is a skill to speake & write harmonically: and verses or rime be a kind of Musicall vtterance, by reason of a certaine congruitie in sounds pleasing the eare, though not perchance so exquisitely as the harmonicall concents of the artificial Musicke consisting in strained tunes, as is the vocall Musike, or that of melodious instruments, as Lutes, Harpes, Regals, Records and such like.[91]

In the Baroque epoch, the interaction between the musical and verbal arts opened up in the rhetorical tradition of composing music, practicing the transfer of rhetorical figures as specific symbols into a musical texture, thus creating a narrative of pure instrumental music (i.e. music without any vocal text). During that period a catalogue of musical-rhetorical figures that served as a means of listening to the musical score *in a different way*, to be able to hear (read) certain signs (messages) in the combinations and harmonies of sounds, was defined. "Textualizing" musical material became a convenient manner in order to implement original solutions of a musical composition and opened up the possibility for subjective intentions of the creators to manifest themselves. These were applied not only from the traditionally defined, but also an individual, subjective aspect.

The question of the musical cryptography was actively discussed in the theoretical works of the 15th century. One of the first descriptions of musical cipher was presented in a 15th century manuscript on the subject of medicine, *Tractatus varii medicinales* (Sloane 351, f.15b), which is held at the London British Library. The cipher is a system of five different pitches and each of their five variants (according to duration and so on) (Sams 2016; Shenton 2008: 69–70). However, as early as in the 10th century, in an anonymous treatise titled *Dialogus de musica* (authorship is attributed to Odo of Cluny, c. 878–942) or even in Boethius' *De institutione musica* (6th century) evidence of alphabetical notation can be found, which is used to denote tonal pitch and which practically was used as far back as in ancient Greece (Shenton 2008: 69). Another example of the evidence of musical cryptography and the alphabetical system was presented in 1602 by the founder of modern cryptography, Giovanni Porta (Giambattista della Porta, c. 1535–1615). He published the treatise *De furtivis litterarum notis* (On Secret Notations for Letters, 1563). The chart model developed by Porta was adapted

91 Cited in George Puttenham: *The Arte of English Poesie*, The Second Booke "Of Proportion Poetical", 1589, p. 53.

Figure 18. Cryptographic models according to Porta, Schwenter, Kircher and Schott (the figure reproduced from Tatlow 1991: 103); according to Wilkins and Terzi (the figures from the manuscripts: Wilkins 1694: 144; Terzi 1670: fig. 1 after page 252)

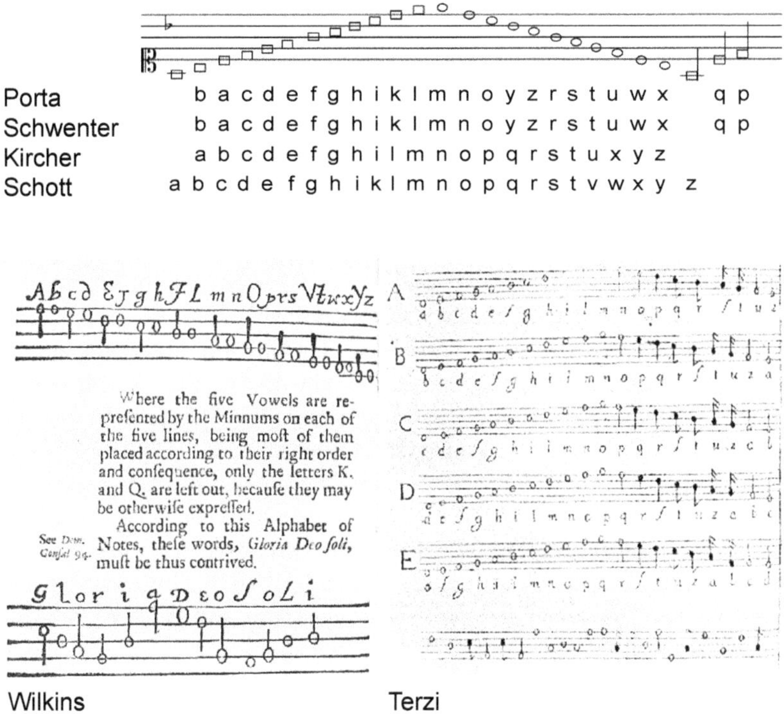

later in the work of Daniel Schwenter, Francis Godwin, Athanasius Kircher, Gaspar Schott, and Johannes Balthasar Friderici[92] (see Figure 18).

Variations of musical cryptography were found in works dedicated to other areas of cryptography as well. In the first encyclopedia of cryptography by Gustavus Selenus (the pseudonym of the book collector Augustus of Brunswick-Lüneburg, 1579–1666), in the sixth section of the ninth book, *Cryptomenytices et Cryp-*

92 Schwenter's treatise *Steganalogia et steganographia*, Nuremberg, c. 1620; Godwin's treatise *The Man in the Moone, or A Discourse of a Voyage Thither*, London, 1638; Kircher's treatise *Musurgia universalis*, Rome, 1650; Schott's treatise *Schola steganographica*, Nuremberg, 1655; Friderici's treatise *Cryptographia, oder geheime Schrifften*, Hamburg, 1685.

tographiae (1624), attention was placed on possibilities of interaction between the musical art and the power of the word. In 1641 the Bishop of Chester John Wilkins (1614–1672) explained a systematization of musical notes with the letters of the alphabet.[93] Wilkins' system became the basis for musical notation ciphers used by the Italian scholar Francesco Lana de Terzi (1631–1687)[94] (see Figure 18) and for the so-called harmonic alphabet by Philip Thicknesse (1719–1792).[95] In 1650, Kircher created a relevant system by transferring the idea of musical cryptography to the orchestra. He offered a code consisting of combinations of four sequential tones and six instruments. For example, the first instrument creates a tone that matches the letter A, two tones match the letter B, and so on (Sams 2016).

These original methods were soon melded into the rich arsenal of music composing practice. One of the first actual examples of musical cryptography is the tonal solmization in Renaissance music. For example, the emphasis on the tone *re* (*d*) in a musical score was compared with the word "King" (Latin *rex*), the tone *sol* (*g*) with "sun" (Latin *solis*). On the other hand, the application of letters in a musical text often associated with the practice of leaving an individual "autograph" within the musical composition. According to one of the most important researchers of musical cryptography, Eric Sams,[96] Georg Philip Telemann (1681–1767), when composing the oratorio *Der Tag des Gerichts* (The Day of Judgment, TWV 6:8, 1762), may have been acquainted with one of the musical cryptographic models. The composition's notes show that the musical tones may be "read" using secret codes, and that information about the behavior and responsibilities of the ambassador and generals would be revealed (Sams 2016).

Based on the variety of examples of musical cryptography from the Renaissance epoch up to 20th century compositions, I would try to systemize the ways of using

93 Wilkins' system was published in his treatise *Mercury, or the Secret and Swift Messenger* (London, 1641).

94 Terzi's treatise *Prodromo all'Arte Maestra* (Brescia, 1670).

95 Thicknesse's treatise *A Treatise on the Art of Deciphering, and of Writing in Cypher: With an Harmonic Alphabet* (place of publishing unknown, 1772).

96 The British musicologist and scholar of Shakespeare's legacy, Eric Sams (1926–2004), is attributed with important and thorough analytical works that research analogies of musical motifs and verbal meaning, the relationship between music and language, and its ties with the composer's creative work of the Romantic period. He is the author of the first discoveries associated with the cipher system typical of Schumann's work, the research on Elgar's cryptographic enigma (Dorabella's code), and has published study about musical ciphers in the work of Brahms.

tonal codes. It was established that in the practice of musical cryptography two directions of tonal coding are evident:

1. the use of individual symbols and signs. The choice of individual letters to represent tones in musical notation, direct correlations between musical tones and letters;
2. the application of a complex system of signs. The musical alphabet as traditionally denoted; individual code systems of letters and tones derived by composers.

1. According to Griffiths, in musical practice there are a variety of ways to assign a letter to a musical tone. For example, according to the French tradition the tone *d* (in Latin tradition *re*) could be "read" as the letter R; *e* (*mi*) could be read as M, *a* (*la*) could be read as L, and so on (Griffiths 2012). The musical motif as a tonal cipher can be made up of individual tone and letter equivalents in musical notation as follows:

the tone *c* is the letter C equivalent
d – D
es – S
e – E
f – F
g – G
a – A
b-flat – B
b – H

Bach's work is a strong example of the personification of his music. The motif *b-flat – a – c – b* (in German convention *b – a – c – h*) can be heard often, especially in the bass part. This is probably not accidental, but the result of the composer's conscious goal to encode his surname into his compositions using musical tones. This four-tone motif as a musical cryptograph had a vital rebirth in the creative compositions of the Romantics. Examples of noteworthy compositions are: Schumann's 6 Fugues on BACH, Op. 60 (1845–6); Liszt's *Präludium und Fuge über den Namen BACH*, S. 260 (1855–6/1869–70); Brahms' Fugue in A-flat minor for organ, WoO 8 (1856); Rimsky-Korsakov's 6 Variations on the Theme BACH, Op. 10 (1878); Reger's Variations and Fugue on a Theme of Johann Sebastian Bach, Op. 81 (1904), etc. Among 20th century composers there was continued interest in this tonal motif. This sound combination is implicated in Schönberg's Variations for Orchestra, Op. 31 (1926–8); Webern's String Quartet, Op. 28 (1936–8); a transposed motif is used in a section of the fugue from the first part of Ives' *3 Pages Sonata* (1905); the motif became the basis of the improvisation in Poulenc's *Valse-improvisation sur le nom de Bach*, FP 62 (1932); it served as a theme in Casella's *Due Ricercari sul nome BACH*, Op. 52 (1932), and appears as a repetitive motif in *Quasi Una Sonata* (Sonata No. 2, 1968) score by Alfred Schnittke (1934–1998). Additionally, in Schnittke's Third Symphony (1981) the motif is installed beside the monograms of other composers. This same motif can be heard in Dallapicolla's piano piece *Quaderno musicale di Annalibera* (1952) and his cantata *Canti di liberazione* (1951–5). The musical monogram BACH can be found in various Arvo Pärt's compositions as well.

Robert Schumann's (1810–1856) work could also be called an encyclopedic illustration of personified musical cryptography. In his musical compositions for piano the codes *SCHA*, *CAA*, *ASCH*, *ABEGG* are incorporated and are full of the *e – f* combination, just like in Eusebius and Florestan analogies. It is believed that Schumann's *Abegg Variations*, Op. 1 (1830) is encoded with a musical dedication to the Countess Pauline von Abegg. In the cycle *Carnaval*, Op. 9 (1834–5) the theme *e-flat – c – b – a* (in German convention *es – c – h – a*) manifests itself constantly, creating the tonal cipher for the composer's surname *SCHumAnn*. His wife, Clara Schumann's (Wieck) name is tonally encrypted through the tonal motif *c – a – a*, *ClArA*.

Besides Schumann another composer of the Romantic era who is often mentioned and who was interested in, and practically applied the cryptograph technique in his work, was Johannes Brahms (1833–1897). Based on biographical facts, it is believed that Brahms in his Sextet for strings, Op. 36, as a farewell to his friend, Agathe von Siebold, *AGAtHE*, created what is referred to as the Agathe motif, *a – g – a – b – e* (in German convention *a – g – a – h – e*). In the Fugue in A-flat minor for organ, in tones *b-flat – a – b – e-flat* (in German convention *b – a – h – es*), he encoded his own name *BrAHmS*. Eric Sams has established that in Brahm's music the appearance of the motif *g-sharp – e – a* (in German convention *gis – e – a*) can be "read" as the tonal transcription of Gisela von Arnim's name *GisELa* (Sams 1971: 329–30).

The compositions of Edward Elgar (1857–1934) are exceptional for the richness and variety of their musical cryptographies too.[97] For example, in 1885 the composer created his *Allegretto (Duet on a theme of 5 notes – GEDGE)* for the Gedge sisters duet. The composer inserted the sisters' surname not only into the

97 Edward Elgar is indicated as the most charismatic of composers who delved into musical cryptography. There are numerous codes and alphabetic messages in his letters and in the manuscripts of his musical compositions. For example, the famous encoded letter from Elgar to Dora Penny (July 1897), referred to as the Dorabella cipher, to this day is difficult to decipher.

composition's title, but in its melody as well (sound motif *g – e – d – g – e*). It is believed that Elgar encrypted the surnames of his music critics into the demon chorus of his oratorio *The Dream of Gerontius* (1900). But the most interesting example is *Enigma Variations* (1899), which is full of secret puzzling. Letter codes are written into the titles of the variations; in those codes the composer hid his name and the names of his friends (more see: Jones 2004, Sams 2016).[98]

It is possible to find certain musical tones linked with letters as a personal autograph in 20th century music as well, for example Russian composer Dmitri Shostakovich's (1906–1975) code. The monogram of the German version of his name *DSCH, d – es – c – h* (English *d – e-flat – c – b*) is incorporated not only into the author's own work (e.g., in his String Quartet No. 8, symphonies No. 10 & 15, Violin Concerto No. 1, and Cello Concerto No. 1 and so on), but inspired other 20th century composers to encode Shostakovich's name in their own music. The tonal motif can be heard in Benjamin Britten's (1913–1976) celebratory cantata *Rejoice in the Lamb* (1943) and in the opera *The Rape of Lucretia* (1946); the motif is widely used in Schnittke's *Prelude in Memoriam Dmitri Shostakovich* (1975); the Russian composer Edison Denisov (1929–1996) dedicated the compositions *DSCH* (1969) and his sonata for saxophone (1970) to Shostakovich, his mentor; the motif can be heard in Dmitri Smirnov's (born 1948) antiphon *DSCH* for two violins (1999).[99]

The other influential monogram, which is widely found in musical texts of the 20th century, is dedicated to Paul Sacher.[100] On the occasion of his 70th birthday composers from 12 different countries were invited to create a piece for cello using

---

98 Among the examples of confirmed musical cryptographs in the compositions of 19th century composers' works are the following: it is believed that Alexander Glazunov "recorded" his pet's name into his *Piano suite on the name Sascha*, Op. 2 (1883); César Cui's First scherzo, Op. 1 (1857), is based on two tonal motifs, *b-flat – a – b-flat – e – g* (German *b – a – b – e – g*) and *c–c*, in which the composer's wife's maiden name *BAmBErG* is encrypted along with his one initials *CC*; in Bedřich Smetana's work one may find the autograph *BS*: *b-flat – e-flat* (or *b – es*) and so on.

99 According to Smirnov, in the Third piano trio *Tri-o-tri* (2005), in the main tonal motif *d – es – e – f – a – h* (or *d – e-flat – e – f – a – b*) he encoded his own initials, and the first letter of his family members: his wife, Elena Firsova, his daughter, Alissa, and his son, Philip.

100 Paul Sacher (1906–1999), a conductor, supporter of contemporary music, member of a variety of committees and institutions founded the famous Zurich Music College (*Collegium Musicum Zurich*). The Basel Chamber Orchestra, which he founded, has comissioned and performed many new compositions (for example, the premiers of Bartók, Hindemith, Stravinsky, Martinů, Honegger, Berio, Carter, Halffter, Henze, Holliger, Rihm, etc.).

the *SACHER* theme *es – a – c – h – e – d* (or in the English convention *e-flat – a – c – b – e – d*). Among them were the composers Holliger, Berio, Lutosławski, Boulez, Dutilleux, Britten, Halffter, Ginastera.

A few special examples of autographs in 20th century music, among others, are the following cases in Alban Berg's work. In the Chamber Concerto (1923–5) he encoded the surnames of three composers from The Second Viennese School: *SCHBEG* (Schönberg), *EBE* (Webern) and *BEG* (Berg). His *Lyric Suite* (1925–6) is made up of motifs from the German monograms *a – b* (or English *a – b-flat*) and *h – f* (or *b – f*) that encode the composer's initials as well as the name of his beloved, Hanna Fuchs-Robettin (1896–1964).

When discussing Lithuanian music, examples of cryptographs are especially apparent in the creative work of Čiurlionis. Implicated in the theme of his *Sefaa Esec* variations (VL 258, 1904) is Stefania Leskiewicz's cryptographic code *StEFAniA lESkiEwiCz*. He wrote into his variations *Besacas* (VL 265, 1904–5) a tonal dedication to the composer's friend, the artist, Bolesław Czarkowski – *BolESłAw CzArkowSki*. According to Darius Kučinskas, from the Polish transcription of the composer's name – *MIkołAj konStAnty CzurlAniS* – the tonal combination *e – as – a – c – as* (or in English *e – a-flat – a – c – a-flat*) is encoded in the tonal structures of the *Easacas* variations (1906) (Kučinskas 2004: 92, 108). Also noted are Balys Dvarionas' (1904–1972) manipulations of the tonal code *a – es – b – d* (or *a – e-flat – b-flat – d*), in which the composer gave meaning to the initials of his name (*BD*) and the name of his wife, Aldona Smilgaitė (*AS*).

2. Cryptography in a musical composition can be put together by continually defining a system of equivalents between tones and letters. For example, the musical alphabet, which can be traditionally established and operated on in the practice of musical composition (one of those, which was offered in the work of Renaissance writers, can be seen in Figure 18) or in the system developed by the composer himself. One of the last examples of such a system was compiled and used practically by Michael Haydn (1737–1806). This particularly perfect cipher system was designed to accommodate all 28 letters in the German alphabet. Unfortunately, it was not widely known among his contemporaries (Shenton 2008: 72; Sams 2016).

Possibilities for refined secret codes using a cryptographic musical alphabetical system was revived at the beginning of the 20th century. One of the first cryptographic coders of this era was Maurice Ravel (1875–1937). While composing *Menuet sur le nom d'Haydn* for piano in 1909 in honor of the 100-year anniversary of Joseph Haydn's death, in his composition the French composer encoded Haydn's name using a 7 x 4 system (see Figure 19). Ravel applied this system in his 1922 musical composition, which uses cryptographic code to honor the composer Gabriel Fauré (1845–1924). Two decades later, Arnold Bax (1883–1953) tonally

Figure 19. Ravel, the first measures of Menuet sur le nom d'Haydn *for piano (1909)*

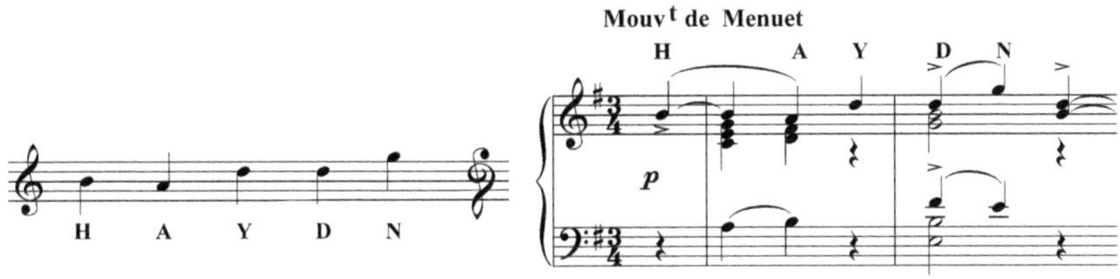

encoded Fauré's name in his composition *Variations on the name Gabriel Fauré* (1949), using the very same cryptographic system by Ravel.

Another common cryptographic system of the 20th century was associated with the name of Albert Roussel (1869–1937), whose name in 1929, on the occasion of his birthday, was inserted in the compositions by Poulenc, Honegger, Milhaud, and Ibert. Poulenc used the 8 x 3 system to write down his cryptographic music. Honegger's practiced cryptographic system is similar to Michael Haydn's version.

The widespread theory of musical cryptography in the musical panorama of the 20th century was applied practically by Messiaen. Most of the music composed by this composer had a paramusical meaning for him. He sought to create a musical language in which it would be possible to communicate and used the French term *langage communicable* (Shenton 2008: 35, 69). The tool of musical communication came into use when this French composer created his system of letter transcripts into musical tones. Messiaen used this method in his compositions *Des canyons aux étoiles...* (1971–4) and his piece for organ *Méditations sur la mystère de la Sainte Trinité* (1969). In the latter composition Messiaen encoded citations in French from St. Thomas Aquinas' *Summa theologica*. In his notes the composer describes in detail how he used the system of the letter alphabet, having chosen the German musical convention *a*, *b*, *c*, *d*, *e*, *f*, *g*, and *h*. He matched the other letters by using phonetic groups. Every letter in this alphabet represents not only a tonal pitch, but also a defined duration.[101]

Among the authors of other original cryptographic systems, it is important to mention the work of the Russian composer Dmitri Smirnov. In his composition for a string quartet and bells *Канон памяти Стравинского (Deum de Deo)* (Canon in Memory of Stravinsky, 1998/2001)[102] he encrypted Igor Stravinsky's

---

101 For a comprehensive study on Messiaen's cryptography see a study by Andrew Shenton (Shenton 2008).

102 The piece is the ninth part of Smirnov's Mass, Op. 105 (1998). The arrangement for string quartet and bells was completed in 2001.

name using the original cryptographic system, which was also used for the compositional process of *Metaplasm 1* (2002). In the composition *The Guardians of Space* (1994) Smirnov experimented with transposing a literary text in English into the space of music. To that end he created a system of letters to match not separate tones, but musical intervals. While studying the cycle *Two fragments* for double bass (1976/98), it became apparent that in the second piece, in the first five tones, using his own cryptographic system, Smirnov encoded his wife, Elena's, name.

In the panorama of Lithuanian music, one aspect of the system of musical cryptography was found in the manuscripts of Čiurlionis. He used a personal cryptographic cipher of the Polish alphabet and musical tones to encode his own name into the draft of his piece *Composition* (see Figure 20). Among Lithuanian composers active in the last decade, Marius Baranauskas (born 1978) adapted his own original system of equivalents of letters and musical elements to his composition for orchestra *Talking* (2002) as well as to later compositions.

Figure 20. Fragments from Čiurlionis' 1906 manuscripts. Above: the musical alphabet (stored in the archive of the M. K. Čiurlionis National Museum of Art, Čm–21, page 260); below: the sketch for unimplemented Composition *for piano (stored in the archive of the M. K. Čiurlionis National Museum of Art, Čm–6, page 00411)*

Application of the Morse Alphabet

The use of Morse code in the creative process may also be attributed to musical cryptography.[103] Almost a century after this artificial alphabet was created, 20th century composers became interested in Morse Code because of its variety of possibilities for the manipulations of musical rhythm. For example, Boulez made use of the organization of rhythmic motion within Morse Code to write a dedication to his friend Paul Sacher into his composition for seven cellos, *Messagesquisse* (1976). His friend's surname reappears over and over again in Morse Code in the rhythmic picture of the different cello parts. Then, in measure 118, it bursts forth in all its glory in a score for all six cellos.

According to the German composer of experimental music and performances Gerhard Stäbler (born 1949), in his composition *O Muro* for voice and two or three percussions (1992), the Morse Code system determines the organization

103 The Morse Code was created in the early 19th century. It quickly became a universal language in a variety of spheres. The author of the code that is used today is Friedrich Clemens Gerke (1801–1888). He sent a coded message from Hamburg to Cuxhaven in 1848.
The alphabet is based on the combination of two elements – the dot and the dash (in fact, three elements, including the space). They express individual elements of the rhythm using original digital patterns. There are two letters that have one element: E and T; 2 elements – 4 letters (A, I, M, N); 3–8 (D, G, K, O, R, S, U, W); and 4–16 (B, C, F, F, J, L, P, Q, V, X, Y, Z, Ä, Ö, Ü, Ch). The principles of the alphabet:
- the duration of one dash is equal to three dots;
- the space between one letter's elements is equal to one dot;
- the space between letters is equal to three dots (one dash);
- words are separated by a duration equal to the length of seven dots.

A	.-	M	--	Y	-.--	6	-....
B	-...	N	-.	Z	--..	7	--...
C	-.-.	O	---	Ä	.-.-	8	---..
D	-..	P	.--.	Ö	---.	9	----.
E	.	Q	--.-	Ü	..--	.	.-.-.-
F	..-.	R	.-.	Ch	----	,	--..--
G	--.	S	...	0	-----	?	..--..
H	....	T	-	1	.----	!	..--.
I	..	U	..-	2	..---	:	---...
J	.---	V	...-	3	...--	“	.-..-.
K	-.-	W	.--	4	....-	‘	.----.
L	.-..	X	-..-	5	.....	=	-...-

Figure 21. Smirnov, pieces for piano from the cycle Ciphers: *No. 1* Invention (Morse-Bach) *and No. 3* Morse-Music. *Rhythmic analysis according to the codes of the Morse alphabet*

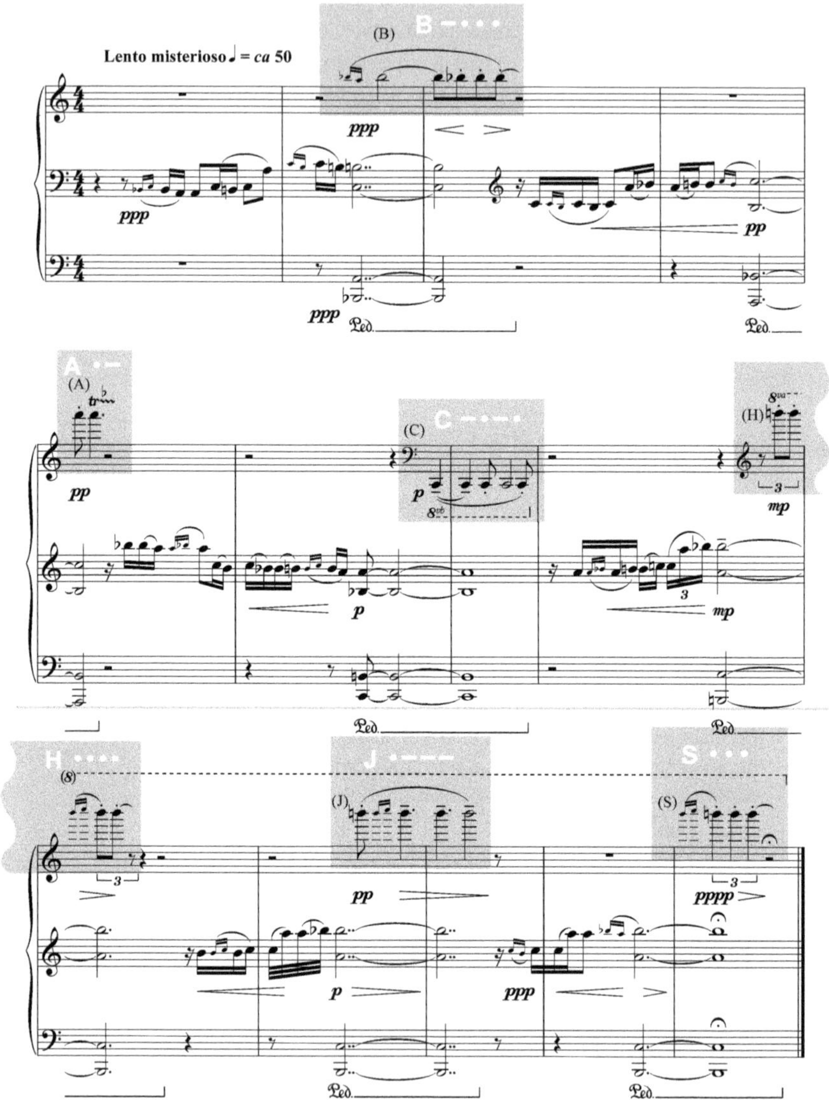

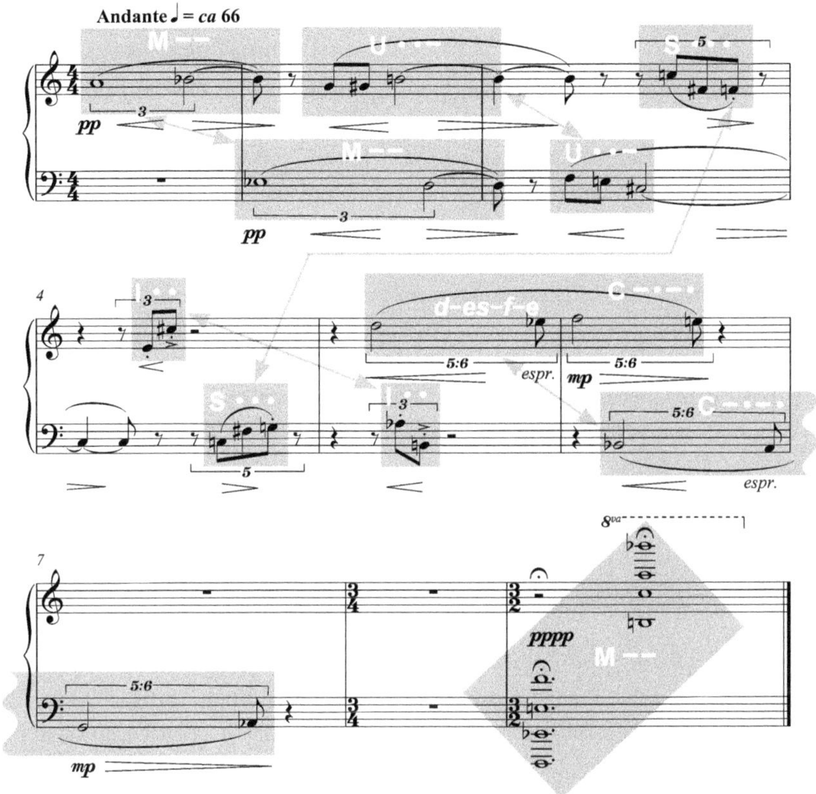
Andante ♩ = ca 66
M
U
S
I
C
d-es-f-e
pp
mp
pppp
espr.
8va
3
5
5:6

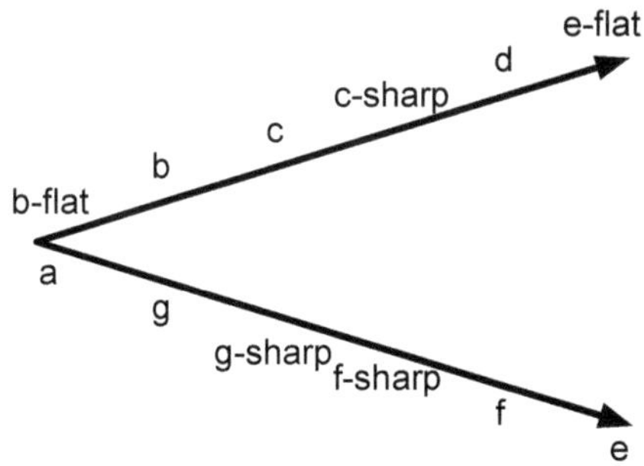
e-flat
d
c-sharp
c
b
b-flat
a
g
g-sharp
f-sharp
f
e

of rhythm within the piece. In this piece, the composer transposed the lyrics of Pedro Tierra's poem *Rosa*, using the symbols of Morse Code. In this manner, he recreated the literary text in the rhythmic material (Ehrler 2000: 15).

Smirnov argues that the rhythmic accompaniment in his *Elegy* for cello (1997) creates an irregular repetition of the great octave *C* (*C2* in scientific notation). According to elements of Morse Code, the rhythmic motion expresses the name of his teacher, the Russian composer Edison Denisov. An especially interesting use of Morse Code in Smirnov's composition is his bagatelles *Ciphers*, Op. 143 (2005). In this cycle he adapts various possibilities of encoding tones. He uses not only Morse Code, but manipulations of cryptographic code and of numerical implications (for example, Fibonacci numbers).

The use of Morse Code is illustrated in the first piece, *Invention (Morse-Bach)*, and in the third piece, *Morse-Music*. Within the bagatelles' rhythmic material he incorporates the verbal codes "Bach J S" and "Music." Additionally, in the first piece the tonal fabric is made up of only a four-tone motif, *b-flat – a – c – b* (in German *b – a – c – h*) and its permutations. In the third bagatelle, which is composed as a strict rhythmic canon, a tonal construction is based on a twelve-tone cone-shaped series. The original of the series is played by the right hand, while the left hand introduces the inversion and the rhythmic canon:

a – b-flat – g – g-sharp – b – c – f-sharp – f – e – c-sharp – d – e-flat

It has been noted that the composer used one more symbol in the third bagatelle. It is the four-letter autobiographical tonal motif *d – e-flat – f – e* that appears in the score on the plane of the letter C). The tones of this motif may be interpreted as the equivalents of the composer's name (*d – e-flat*, or *d – es*, as Dmitri Smirnov) and the name of his wife (*f – e* as Elena Firsova) (see Figure 21).

4. *Musica Mathematica* in Practice: Aspects of Analysis

In the 20th–21st centuries, the practice of the constructive composing of music is a diverse result in which mathematical manipulations of earlier epochs and modern innovations coexist. Therefore, a wide range of analytical procedures are applied in the analysis of a contemporary musical score, beginning with the universal methods, for example of Renaissance, Baroque or Classicism, to a search for new approaches. Apart from the traditional approach to pitch, rhythmic pattern, musical time, duration, dynamics, timbre, and sonic parameters of a musical composition, we may analyze the arsenal of a sonic expression, which is obviously supplemented with a variety of articulation, scales of dynamics, groups, layers, polytempo and other aspects.

Purely *mathematical procedures*, which create the possibility to determine and derive, for example, primary numbers, which generate a certain meaning or organize the structure of a composition, are proposed to be employed in the investigations. Therefore: deriving the digital root (the compound number is reduced to a single digit), diminution and augmentation actions, etc. These mathematical actions, such as the reduction of initial compound numeral to its root – a single digit result. For example, in Christian semantics the number 10 can be adapted to the results of its digital augmentation – 10, 30, 100, 1,000, and so on. Therefore, 1,000 can be interpreted as number 10 with exaggerated significance, because 1,000 is the cube of 10. In other words, it is a strengthening of the symbol of the Holy Trinity. In Christian numerology multiplying a number by three is an especially significant action. By this logic, the number 9 becomes a particularly sacred number, as it is the result of multiplying 3 x 3.

Having chosen *the parameter of the pitch* as the subject of analysis of a musical composition, the following can be analyzed:

- the number of tones on the scale;
- the calculation of tone attacks of the chosen structural segments according to the *Keystroke* method (*Keystroke* is an indicator of the number of tones actually struck/attacked by the performer). The account of key strokes is made in different parameters of a musical composition – in separate sections of the form, themes, sub-themes, lines of a single vocal part, chord, phrase, and other compositionally significant fragments;
- the row of tones can be reduced to a certain sequence (cluster indicator);

- Allen Forte's set theory – a method of recording the pitch in terms of numbers is applied: likening the pitch to the sequence of whole numbers from 0 to 11 (*c* – 0, *c-sharp* – 1, etc.).

Investigations of *the tonal parameter*. The number of key signatures becomes a numerical emblem of specific tonality in a tonal musical composition. For example, tonalities of few key signatures appearing in the music of the Baroque period were most often related to the numerical equivalents of the sacral symbols of significance to Christian numerology: e.g., E-flat major with its 3 flats is related to the sphere of Angels, B-flat minor – to Christ's five wounds as an allegory of Christ's suffering. Also, the method of establishing an additional numerical cipher of tonality is proposed to be used when investigating the tonal structure of a composition: each tonality is equated with a specific number, which is obtained by counting it in the order of sequence in semitones from *c* upwards. For example, F-sharp major and F-sharp minor is represented by number 7, because *f-sharp* is the seventh semitone from *c*.

In modern music, however, it is the affinity between the specific sonic centers, for example the two prevailing sounds, rather than tonalities, that are expressed on a numerical basis, in terms of a numerical formula. The following can be analyzed:

- indicators of a twelve-tone series. They can be calculated in dodecaphonic music;
- the phenomenon of integral (total) serialism. The constructivism of all musical parameters (duration, dynamics, pitch, articulation) choosing the number uniting these parameters can be analyzed;
- the sonic effects created by the sonorism can also be rationally substantiated. For example, the transition from one sonic block to another is expressed in terms of quantitative relations of the sounds or durations forming those blocks.

When taking account of *the vertical* of the musical score, it is necessary to pay attention to the following:

- the number of lines, parts and layers that form the score;
- tendencies for the permanence or changeability of vertical indicators. For example, the duration of the same number of parts (lines) in the vertical; setting the number that organizes the chord vertical; what number of tones forms the chord; how that number of tones changes, etc.

Investigations of *the parameter of time of music* encompass such elements of the form of a musical composition as sections, bars, logic of the duration of larger and smaller rhythmic structures, the formation of the pattern of rhythm, the composition of bar groups, and establishment of the relations between different time signatures. In this case, the numerical relations between the parts or sections of the composition are investigated. Possibilities of dividing the composition into smaller sections or phrases are established. Volumes of structural sections, fragments, arrangement of thematic models from the point of view of the bar groups are calculated.

When analyzing models of rhythmic relations, the following procedures can be carried out:

- the quantity of certain formulas is summed;
- regularities and patterns of rhythmic formulas are established. For instance, how many times the composer repeats the same rhythmic figure, how many rhythmic values form a compositionally significant rhythmic complex, what duration of the rhythmic model is measuring it in terms of a chosen time unit, what internal peculiarities of the structure of a rhythmic model are;
- the impact of the sequence of numbers on the rhythmic process (including the formation of rhythmic formulas, changes in the time signature/meter, grouping of bars etc.) is investigated.

The theme or a thematic (nuclear) motif, a subject or melody is the most important image of the composition and is a compositional-constructive element. Different investigations of the above-discussed musical parameters are applied to the musical theme's numerological interpretation in a complex way. Also, the numerical expression of the theme can be analyzed from the following aspects:

- the calculations of the theme or the number of sub-themes, its alterations (original version, transposition, ètc.);
- the volume in bars (half bars) or the length in other units of time (e.g., seconds);
- rhythmic structure of the theme;
- the number of tones forming the theme;
- peculiarities of the tone scale.[104]

The establishment of *mono-numerical factor*. The analysis of a musical composition carried out convincingly, the substantiation and effectiveness of the operations applied, is achieved when one manages to find and purify the primary constructive element, the particular number, as the initial structure organizing the sounds, appearing in different parameters and thus uniting the

---

104 A numerical expression of the ratio of black to white keys is also proposed in piano music; numerical relationships in dividing the sounds of the theme between the right and the left hand can be calculated.

musical material. This may bring out a base number (as the mono-numerical or mono-constructive principle) that serves both as a tool of composing and as a deliberate notional precondition of the creation of the opus itself. As an example, I provide the notes on mono-numerical analysis of the piano compositions by Bach and Webern. The Fugue in F-sharp minor (WTC Vol. 1, BWV 859) shows the extensive manifestation of number 7 and its numerical formula 3 + 4 or 3/4, for example:

- *f-sharp* is the seventh pitch calculating from *c*, and the very middle of octave. Therefore, the fugue is recorded as No. 7 in the first volume of The Well-Tempered Clavier
- the time signature 6/4 refers to the dual expression of 3/4
- the volume of fugue subject in half bars is 7 or in quarter notes – 21 (7 x 3)
- the number of subject tones reduced to a certain sequence (cluster indicator) is 7 (*f-sharp*, *g-sharp*, *a*, *a-sharp*, *b*, *c/b-sharp*, and *c-sharp*)
- the cluster indicator of the answer and countersubject (starting in m. 4) is 7 tones each as well
- the ratio of black to white keys of the subject shows the relationship of 3/4: 4 black keys (*f-sharp*, *g-sharp*, *a-sharp* and *c-sharp*) and 3 white keys (*a*, *b*, and *c/b-sharp*)
- the ratio of black to white keys of the answer and countersubject is the same manifestation of 4 black and 3 white keys
- the rhythm of countersubject is a combination of eighth, quarter and half notes, in total there are 14 eighth notes
- in the first 7 bars there are 49 eighth notes (7 x 7) and 7 quarter notes
- in the first 7 bars *f-sharp* appears 7 times
- formula 3/4 manifests in the construction of subject that consists of two ascending trichords (*f-sharp – g-sharp – a* and *g-sharp – a-sharp – b*) and one tetrachord (*g-sharp – a-sharp – b-sharp – c-sharp*)
- the exposition of fugue in half bars is 34 and thus refers to the expression of 3/4
- harmonies of F-sharp minor and C-sharp minor dominate the musical material, accordingly their key signatures (3 and 4 sharps) refer to the formula ¾
- the whole fugue consists of 16 structural elements: 9 statements of the subject, 6 episodes and coda; the digital root of the sum of these elements is 7 (9 + 6 + 1 = 16 → 1 + 6 = 7)
- 7 statements of the subject are presented in the original shape, other 2 provide the inversion of the subject

The entire construction of the first piece from Webern's cycle, *Drei kleine Stücke* for cello and piano, Op. 11 (1914), is subordinated by the number 8:

- 8 measures
- 8 different durations:

  𝅘𝅥𝅯(triplet) 𝅘𝅥𝅯 𝅘𝅥𝅯. 𝅘𝅥𝅮‿𝅘𝅥𝅯 𝅘𝅥𝅮. 𝅘𝅥 𝅘𝅥..
- 8 changes of tempo:
  a tempo, rit., a tempo, accel., rit., a tempo, rit., a tempo
- 8 signatures of dynamics:
  pppp, ppp, pp, p, mp, mf, sfp, f
- the eighth note is chosen for the time signature 6/8
- the piece is composed of two sections and the cello plays eight different tones in each of them
- the highest pitch *e-flat*^{3} in the piano part appears as the 8th eighth note starting from the beginning
- besides, the structure of the piece shows the manifestation of Golden Ratio, because the lowest pitch *E-flat* in the piano part appears in the sixth measure, or 33rd eighth note, that is the Golden Ratio of the composition taking into account the total duration of 54 eight notes (according to the formula *n* x 0.618, that is 54 x 0.618 = 33.372)

Part 2
The Renewal of Mathematical Techniques in Musical Compositions of the 20th and 21st Centuries

In this part I explore the polyhedral phenomenon of *musica mathematica* in 20th–21st century musical compositions. One might say that the compendium of *musica mathematica* of earlier epochs engenders a peculiar eclectic combination, one that employs a combination of a variety of different constructive manipulations. When analyzing the features of the creative process, one encounters models of additivity, progressions and symmetry, which have traditionally established themselves in the practice of composing music. In contemporary musical composition the traditions of applying antique proportions, Kabbalistic numbers, Christian numerology, and numericalized semantics have been revived. I also would not be incorrect in noting that this variety of tools used in composing music helps determine contemporary composers' aspirations for individuality and exclusiveness, which dominates our modern worldview and which manifests itself in the especially personified intentions of composers.

It would seem that never before has the creative space of music been filled with the search for original results as it is today. Opuses of contemporary music seek to "hide" an original special way of "deciphering", rather than applying a universally determined numerical interpretation. Most often it is difficult, or even impossible, to establish a specific semantic or logical code without the author's interference (his own testimony, or the like). This is because in most cases a contemporary musical composition is a one-off realization of a certain idea; a specific model is applied in the case of that musical composition only. Nonetheless, it is possible to systematize and generalize the renewal of mathematical techniques in 20th–21st century music by making use of the dichotomy of constructivism and semantics peculiar to the multiplicity of the centuries-old phenomenon of *musica mathematica*. Thus the spread of the traditions of earlier epochs can be divided into two categories: the first, formal-constructive, and the second, semantic-symbolic ways of introducing numeration into musical scores. Research has shown that these two trends may function separately or as a synthetic interaction.

1. Constructive Aspects of Music Composition

The process of the application of a constructive tool to organize sounds is of a purely technological nature, because manipulations with numbers, like structural models, are technical. The rational nature of numbers emerges in different parameters of a musical composition. For example, it determines the general organization of pitches, or the peculiarities of the original series, and the internal structure of the melody. It is inserted into the parameter of the duration of the whole composition, its sections, or even smaller sonic units. This has an effect on the rhythmic relief, and so on. In the case of the constructive composing of music, the numerical formula chosen, a certain progression or another mathematical model, is merely a "tool" of the creative act in the composer's laboratory and does not render any notional underlying implication to the final result. Such cases are further presented in analytical observations of different scores that are grouped according to general creative intentions. They are as follows:

- implications of numerical proportions and progressions, including the use of antique proportions, the Fibonacci and other progressions;
- the expression of algorithms of symmetry, confrontation of symmetry and asymmetry;
- the renewal of polytempo, polyrhythm, polymeter, as constructive Renaissance principles of composing music;
- the application of combinatorial operations with the help of diverse transformational models (permutation, rotation, etc.).

1.1. The Implications of Numerical Proportions and Progressions in Music

The first part of the monograph highlighted a retrospective examination of the reasons why artists have been intrigued by the implications of numerical proportions and progressions. Universal codes of logic are hidden within the relationship between numbers. These codes of logic are not senseless in any elementary sense, but are the embodiment of an elegantly refined inner order on the basis of universal harmony. It is therefore understandable that these historically entrenched and irrefutable laws were successfully adopted in 20th–21st century music. Contemporary music has been widely manipulated with geometric, arithmetic, and harmonic proportions defined in ancient Greece, and the progressions that are based on them. The sound space is structured with the Fibonacci numbers, Lucas series, and the *Série Évangélique*. We also encounter other mathematically defined sequences, such as prime (Eratosthenes) number sequence, the numbers of the mathematician Marin Mersenne, etc.

1.1.1. The Number Proportions and Progressions of Antiquity

In ancient times the idea of recording beauty with the assistance of mathematical formulas was raised and remained effective in later epochs, becoming one of the major principles of beauty (or harmony). The arithmetic, geometric, and harmonic proportions and progressions that attracted attention for their logical nature and which are characterized by a consistent "growing" are obvious in 20th–21st century musical compositions. Research has shown that their models are often used to organize the musical rhythm. For example, Charles Ives (1874–1954) transfers the law of arithmetic ratio to the second part of his String Quartet No. 2 (1913–5) by creating a constant rhythmic pattern diminution in the fugue's subject: one quarter note dividing into two eighth notes, then into eighth note triplets, then into four sixteenth notes, and so on:

The law of regular arithmetic sequences leads the rhythmic pattern of the two etudes (No. 1 & 2) of Boris Blacher's (1903–1975) piano cycle *Ornamente*, Op. 37 (1950). Fragments of the following progression can be found in the rhythmic parameter of the design of Etude No. 3: 2–3–4, 3–4–5 and 4–5–6; in Etude No. 7 the length of individual measures in values of eighth notes is composed according to a retrograde and a constantly elongation of arithmetic pattern (see Figure 22).

The American minimalist Tom Johnson (born 1939) in his composition for organ *Mélodie de six notes* (1986) employed a slightly different logic to organize its length parameters. The four sections of the composition are progressively each seven times longer.

Figure 22. Blacher, Etude No. 7. Metrorhythmic design of the composition

An interesting example of arithmetic progression is represented by the work of Lithuanian composer Osvaldas Balakauskas (born 1937). His original method was applied to various compositions. It is made up of three rhythm series: the first is typified by the progressive division of the length of the main component; the second is the progressive enlargement of the main component; and the third is typified by the same rate per unit progressions (Mikėnaitė 2000: 16–7). For example, all of these series were implemented in the rhythmic structure of his sonata for piano *Cascades* (1967).

It can be said that the application of numerical relationships to the rhythmic parameters of a musical composition are especially convenient for the encoded construction of the rhythm itself. However, the numerical progressions are adapted on the plane of melody as well. For example, change in melody regulating geometric progression is inherent in the logic of Johnson's composition *Doublings for double bass* (1980). At the beginning of the composition the melody consists of two tones, and gradually it is enlarged with new tones. The quantity of tones are regularly increased according to a geometric relationship: from the cell of two-tone melody a melody of four tones grows, then 8, 16, 32, and so on.

Olivier Messiaen. *Quatre études de rythme* (1949–1950)

An original and refined technique of composing is distinguished by the French composer Olivier Messiaen. His composition theory makes use of the concept of chromatic durations, which describe a series of rhythmic values based on an arithmetic progression used in his work. The composer applied this technique practically, for example, when he composed the etude *Île de feu II* from his cycle *Quatre études de rythme* (1949–1950)[105] using the twelve-tone series: the chromatic progression of durations is recorded in the header of the etude as a tone-series prototype, formed by the chromatic scale down from *b* to *c*; each tone of the series is one sixteenth note longer than the preceding tone (see Figure 23). In this etude, and later in the symphonic composition *Chronochromie* (1959–60), Messiaen, striving to create diversity from the same musical material, used an original technique of interversions (permutations).[106] In the rhythmic organization of the third etude of the same cycle, *Neumes rythmiques*, we encounter an even more complex, though refined, instance of arithmetic progression – a numerical progression according to the "triple line" principle (see Figure 24).

The arithmetic formula 1 : 2 : 3 is used in Messiaen's second etude *Mode de valeurs et d'intensités* for the organization of three rhythmic series of 12 tones each. Their order is regulated by rhythmic values, which are arithmetically chosen from the preceding series, whereas the geometric link 1 : 2 : 4 creates the relationship between the series (see Figure 25).

Figure 23. Messiaen, Etude Île de feu II. *A chromatic tone-series prototype and sequence of durations*

Figure 24. Messiaen, Etude Neumes rythmiques. *The "triple line" principle*

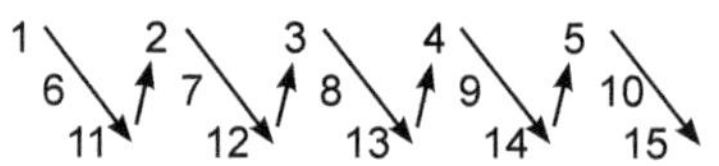

105 The musical composition is comprised of four etudes: No. 1 *Île de Feu I* and No. 4 *Île de Feu II* composed in 1950, and No. 2 *Mode de valeurs et d'intensités* and No. 3 *Neumes rythmiques* composed in 1949.

106 A detailed analysis of the etude *Île de Feu II* is presented in the section "Transformational Elements (Combinatorics, Permutations, Rotations)" in this book, pp. 152–156.

Figure 25. Messiaen, Etude Mode de valeurs et d'intensités. *Organization of rhythmic design and notes of the scale according to arihtmetic and geometric orders*

Snieguolė Dikčiūtė. *The Mystery of Seven Bridges* (1991)

The creative intensions of the Lithuanian composer Snieguolė Dikčiūtė (born 1966) usually combine rational constructivism with a symbolic narrative. This is illustrated in the vocal cycle *Septynių tiltų misterija* (The Mystery of Seven Bridges, 1991).[107] In various parameters of the composition models of geometric and arithmetic progression emerge. For example, all seven parts of the cycle symmetrically employ four rhythmic series; the inner relationship of the series is based on arithmetic logic; the rhythmic representation of each part creates their own geometric relationships. All this is revealed by the rhythmic sketch of all

107 The semantic subtext of Dikčiūtė's composition is analyzed in the section "The Implications of Sacred Numbers" in this book, pp. 181–182.

Figure 26. Dikčiūtė, The Mystery of Seven Bridges. *Arihtmetic and geometric principles of the rhythmic organization*

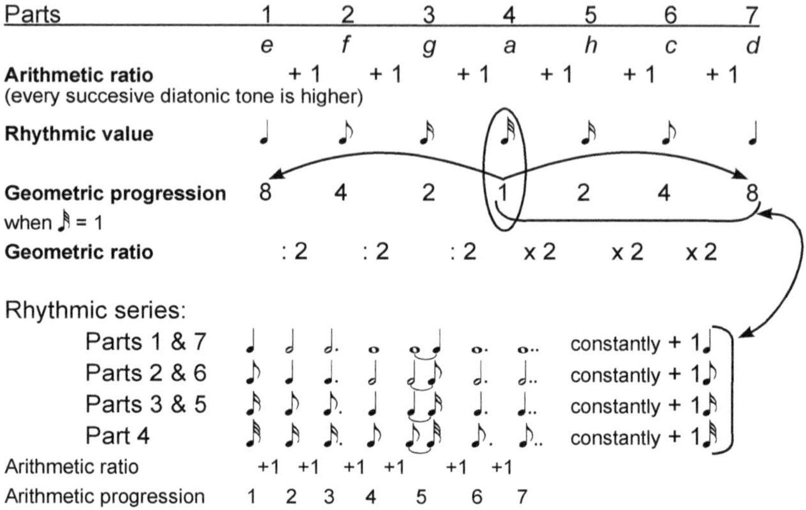

seven parts (see Figure 26). Besides, the logic of arithmetical steps can be seen in the organization of the "white-notes" scale (diatonic scale of white key tones) that unifies the piece. At first the scale and its rotational examples are created from the tone *e*, and later sequentially from *f*, *g*, *a*, *b*, *c*, and *d*.

Steve Reich. *Music for Pieces of Wood* (1973)

The influence of the numeric formulas of Antiquity can be seen in the work of Steve Reich (born 1936). Reich is one of the first practitioners of repetitive music (which has its origin in the 1960s in New York). His composition for percussion *Music for Pieces of Wood* (1973) not only expanded the composer's explorations in the field of percussion timbre,[108] but also displays rational game creating the composition's rhythmic design.

According to Tom Johnson, in the background of Reich's music there is a strict, sometimes obvious, sometimes masked, logic (Johnson 1989: 35–6). This remark can be applied to the composition *Music for Pieces of Wood*. Here the progressive rhythm formulas are manipulated as constructive principles of creation, which

108 The search for percussion timbre is typical of Reich's *Clapping Music* (1972) and two compositions created in 1973: *Music for Mallet Instruments, Voices, and Organ* and *Six Pianos* (in 1986 revised as *Six Marimbas*).

influence the logically polished structure of the piece. The composition is performed with tuned claves by five musicians, producing a four-tone scale in the vertical of the score, *a – b – c-sharp – d-sharp*. Reich expands upon this limited arsenal of tones with an organizational variety of other parameters.

Firstly, the 59-measure composition changes its time signature three times, dividing it into three sections. Within the metric changes one sees a retrograde of the harmonic formula from ancient Greece – 3 : 4 : 6, because time signature in the first part is 6/4, in the second – 4/4, and in the third – 3/4.

Secondly, certain rules of numeric relationships from Antiquity are common to the organization of rhythm. The part of the first musician is rather monotone due to its typical exchanges of eighth notes and eighth-note rests (♪ and 𝄾). However, a refined rhythmic image emerges in the score of other parts (played by the 2nd, 3rd, 4th and 5th musicians), which is unique for its rhythmic formulaic interchanges and organic interlace. Besides, each section's rhythm is modeled on the rhythmic figures from the previous section. Having designed the rhythmic diagram it became apparent that the composition is divided into three sections, and not only to enable new metric directions to be introduced (see Figure 28). The partitions also serve as the borders between changes in rhythmic structure. For example, the first section is made up of two symmetrical pairs of performers: the pair of the 2nd and the 5th musicians, like the 3rd and the 4th musicians, performs the same rhythmic formula and its rotation. The parts of the 2nd and 5th musicians implement the common rhythmic model in the second and third sections as well. But, the rhythmic motion of the 3rd and 4th musicians does not match, as they use the permutations of the same rhythmic formula differently (see Figure 28).

Thirdly, the diagram of the entrance of instruments reveals the logic in the gradual accession of the 3rd, 4th and 5th musicians: the next performer only comes in after the earlier rhythmic formula has been completed (see Figure 29). And fourthly, having analyzed how the composer weaves the entire rhythmic formula into the composition, it became clear that inside the formula an arithmetic progression was applied: one tone at a time was introduced into each new measure, slowly completing the rhythmic pattern. This is typical of the 3rd, 4th and 5th performers, as can be seen in Figure 30.

Figure 27. Reich, Music for Pieces of Wood. *Vertical arrangement of four-tone scale, mm. 18–20*

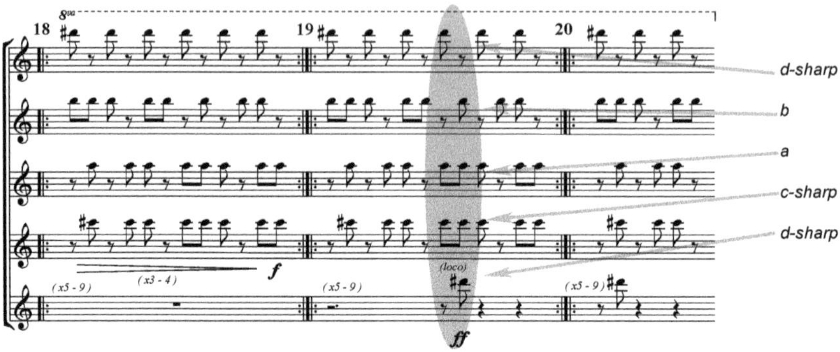

Figure 28. Reich, Music for Pieces of Wood. *Rhythmic design of the three sections*

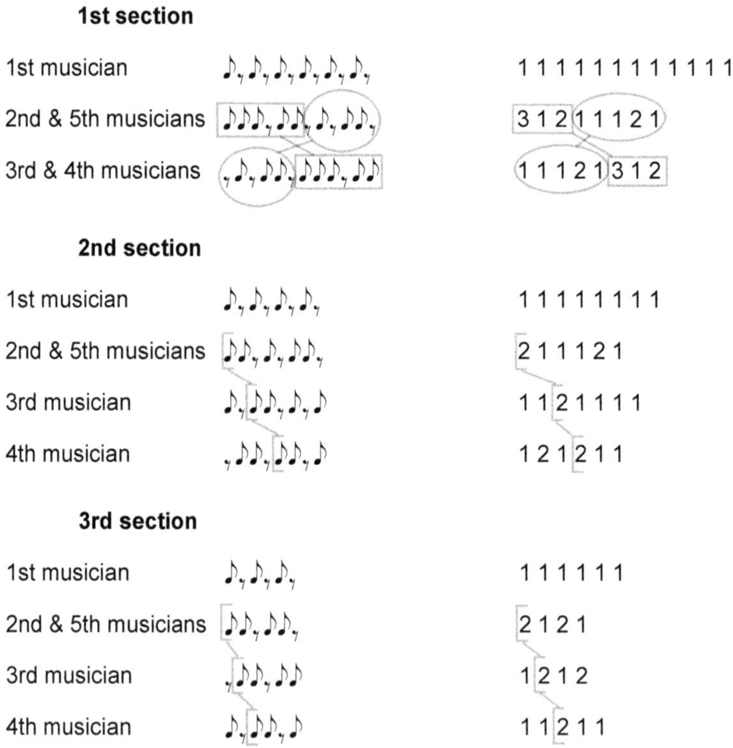

Figure 29. Reich, Music for Pieces of Wood. *Diagram of the entrance of musicians*

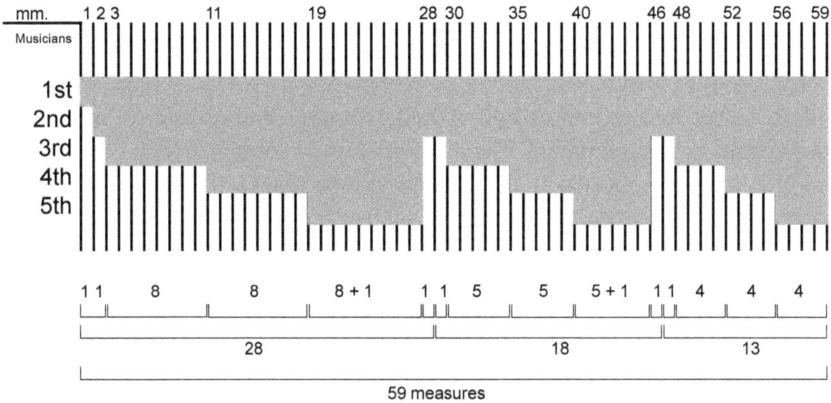

Figure 30. Reich, Music for Pieces of Wood. *Constant "growth" of rhythmic formula in the part of the 4th performer, mm. 11–8*

Siegfried Thiele. *Proportionen* (1971)

The unique aspect of the German composer Siegfried Thiele's (born 1934) work is the harmony between the liveliness of his music and constructive counting, which is applied to his dodecaphonic piece *Proportionen* for oboe, cello, and piano (1971). In the structure of this piece the composer's named rules of numerical proportions become apparent. The manifestation of constructivism is witnessed not only in the composition's title, but also in the notes with certain numerical proportion for the score. Beside the titles of the three parts, *Subjectum*, *Interludium*, and *Toccatina toccatissima*, the composer indicated the following numerical formulas 5 : 4 : 3, 1 : 1 and 1 : 2 : 3. In his notes, Thiele confirmed that the relationship 5 : 4 : 3 relies on the parameters of the first part (groups of measures and the time signature division); the proportion 1 : 1 organizes the instrumentation of the second part (the melodic process of the oboe and cello); and the proportion 1 : 2 : 3 is applied to harmonize the instruments in the third part.[109]

It has been established that these number formulas influence the organization of the structure of three twelve-tone series. The composer increased the music intervals according to the numerical formulas. Thus, the diagram of tone-series acquired a cone-shaped form (as seen in Figure 31). For example, in the first part the basis of music material is a series that is expanded symmetrically up and down and with the typical progression of intervals from m3 to P4. The digital expression of these intervals' semitones matches with the piece's numerical formula retrograde 3 (m3) : 4 (M3) : 5 (P4). The second and third parts' dodecaphonic series are constructed analogically.

It was also established that the mentioned digital proportions influence the index of meter. For example, the proportion 5 : 4 : 3 in *Subjectum* influences the continual exchanges of four time signatures – 1/4, 3/4, 4/4, and 5/4. Changes in the time signature according to proportional relationships are especially colorful in the second section of *Subjectum* (mm. 53–80) and in the coda (mm. 103–12) (see Figure 32).

---

109 German "Auf die Gliederung des Zeitverlaufs im ersten Satz (Taktgruppen, Taktarten), auf die Instrumentation im zweiten Satz (Anteil der Oboe und des Violoncells am melodischen Geschehen), auf die Stimmigkeit im dritten Satz." Quoted after Siegfried Thiele: *Proportionen für Oboe, Violoncello und Klavier*, VEB Deutscher Verlag für Musik Leipzig, 1974.

Figure 31. Thiele, Proportionen. *A cone-shaped arrangement of three 12-tone series and rhythmic progression*

Figure 32. Thiele, Proportionen, *part 1* Subjectum. *Exchange of time signatures according to the numerical formula 5 : 4 : 3, in mm. 53–80 (2nd section of the part) and mm. 103–12 (3rd section)*

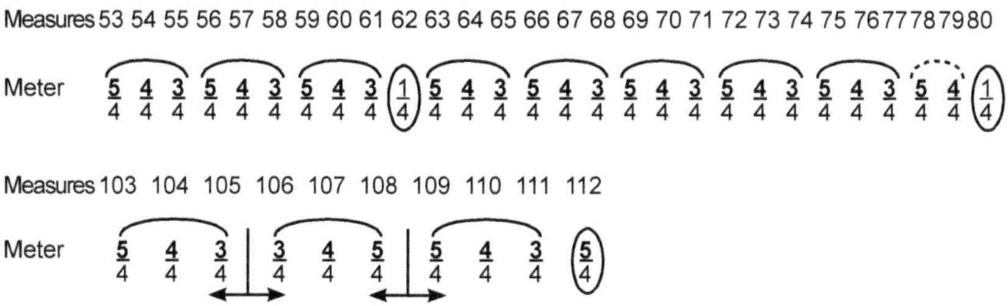

Figure 33. Thiele, Proportionen, *part 1* Subjectum. *Influence of the numerical formula 3 : 4 : 5 to polyrhythmics in mm. 29–30, 32–3, 89–90 and 112*

The influence of digital relationships in the rhythmic organization of *Proportionen* is especially pronounced as the harmony of rhythm between the instruments. For example, in the beginning of the first part, the melody of the oboe is composed of quintuplets (5) of sixteenth notes, while the cello plays the triplets (3) of eighth notes and the piano has a movement of four sixteenth notes. Later this polyrhythmic model changes its vertical arrangement, and then in the final measure of *Subjectum* (m. 112) all the instruments play a dodecaphonic scale, which is rhythmically expanded according to the numerical series 3 : 4 : 5 (a retrograde of 5 : 4 : 3) (see Figure 33).

The second part, *Interludium*, is, in contrast to the two other two parts, a free movement. *Interludium* has no time signatures and no bar lines. The twelve-tone row is composed as an example of integral (total) serialism – every tone characterizes a certain dynamic, articulation, and duration unit (see Figure 34). The numeric formula 1 : 1 in this part can be interpreted as the expression of two equal parts for two instruments, the oboe and the cello.

In the third part, *Toccattina toccatissima*, again, the composition of the trio is manipulated. The compositional process is returned to the construction of digital relationships. Here, the proportion 1 : 2 : 3 influences the transitions of instrumental parts in the vertical. Tendencies were noted when it was counted how many bars and how many instruments play at the same time. The results were transcribed into numeral series. Numerical proportions were also established by analyzing changes in time signature. For example, in mm. 1–36 two digital structures are manipulated at the same time: time signatures are written according to the numerical progression from 2 to 7, whereas the proportional relationship 1 : 2 : 3 designs the groups of measures (see Figure 35). While notating time signatures in mm. 37–68 it becomes clear that there is a symmetrical row of numbers, using the arithmetic progression 2–3–4–5–6–7.

Figure 34. Thiele, Proportionen, *part 2* Interludium. *Example of total serialism at the beginning*

Figure 35. Thiele, Proportionen, *part 3* Toccattina toccatissima. *Manifestation of numerical formulas 1 : 2 : 3 and 2–3–4–5–6–7 in mm. 1–26, 1–36 and 37–68*

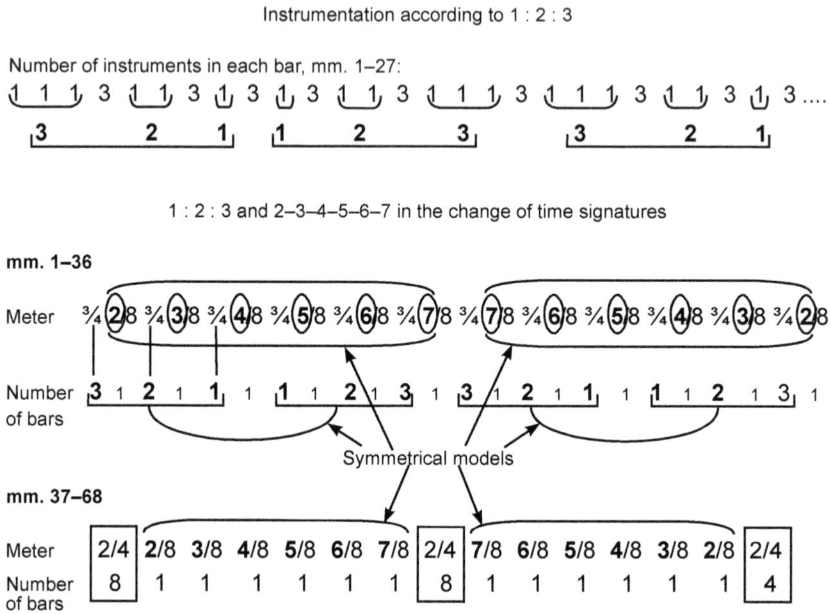

John Cage. *First Construction* (1939)

The abstract numeration in John Cage's[110] (1912–1992) compositions is used as the symbolism of measuring time, for example the silent piece *4'33"* (1952). The title specifies the composition's length that is recorded in minutes and seconds. Other similar music compositions: *4'33"* second version *0'00"* (1962), composition *59½* for any four instruments (1953), *34'44.774"* and *31'57.9864"* for

110 Aleatoric, or the principles of randomness and probability in music are often applied to the compositions of John Cage and are identified with original tonal experiments (music for 12 radios in *Imaginary Landscape No. 4*, 1951), the justification of noise in music or the use of the new aesthetic of silence as one more expression of tone (silent piece *4'33"*, also called, *Silent Prayer*, 1952).
The effect of Cage's ideas are described as "pop art, happenings, multi-media, minimalism, concept art, and contemporary music theater" (citation according to Johnson 1989: 98).

piano (1954), *26'1.1499"* for strings (1953–5) or *27'10.554"* for percussions (1956) and so on. Numbers became the inspiration of Cage's impressive series of compositions, titled *number pieces* as well.[111] However, even earlier, in the 1930s and 1940s, compositions were unique in their expression of Cage's tendency towards constructivism. An especially strict rhythmic structure was supported by the percussion sextet *First Construction* (1939) or in the composition *Imaginary Landscape No. 1* (1939). The latter piece marked not only the composer's early experiments using electronic music,[112] but also his interest in constructing so-called duration blocks when the composition's rhythm is composed according to a model that had been prepared in advance.

Imaginary Landscape No. 1 is composed of a unique organization of measures according to the sequence of numbers 5–5–5–1–5–5–5–2–5–5–5–3–4; groups of five measures are separated by short interludes. The duration of interludes progresses from 1 to 3 measures, and the composition ends on the 4-measure coda. Griffiths, intrigued with the Cage's organizational time principle, identified its prototype in George Antheil's *Ballet mécanique* from 1924 (Griffiths 1981: 11).

Cage soon applied the technical possibilities of the experimental rhythmic structures that were adapted to *Imaginary Landscape No. 1* in his sextet for percussion *First Construction.* Having analyzed this composition, it was noted that groups of measures are organized according to the palindromic number sequence 4–3–2–3–4 (see Figure 36). According to Cage, the idea behind the structural organization of *First Construction* is connected with Erik Satie's (1866–1925) piece *Choses vues à droite et à gauche (sans lunettes)* (1916) in which there is a typical 2–3–4 combination of measures.

---

111 Cage started to create the number pieces in 1987. A total of 43 pieces were composed within five years (e.g., *One* [1], *Two* [2], *Five* [5], *Three* [3], *Four* [4], *Fourteen* [14], *Eight* [8], *Fifty-eight* [58], *103*, *108* and so on). Incidentally, Cage worked on the composition *One* until his death. In total, he produced ten versions of this composition, each for varying timbres. In the final year of his life, he produced the film *One 11*.

112 *Imaginary Landscape No. 1* is often referred to as the earliest example of electronic music. The composition is written to be performed on electronic instruments, plates, and piano. Although it is a rather modest example of live electronic music, because often through the use of tapes consistent sounds were produced or sounds similar to that of sirens like *glissando* (Griffiths 1981: 7).

Figure 36. Cage, First Construction. *Structural diagram of symmetrical macrostructure 4–3–2–3–4*

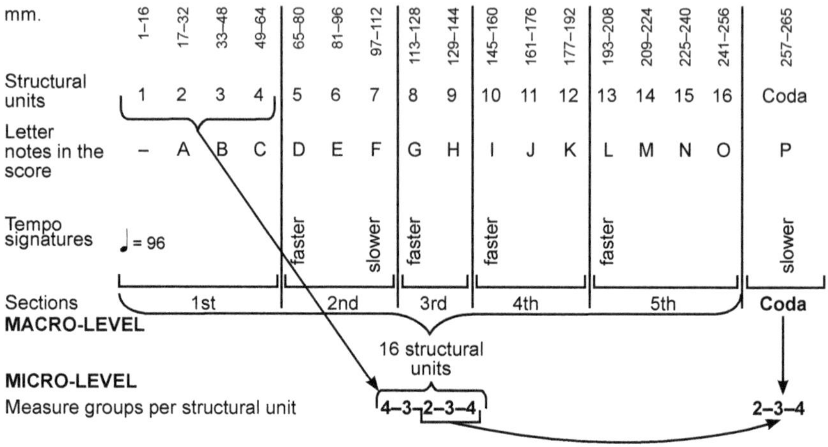

First Construction is composed of 16 units (blocks), which are organized into five sections of 4, 3, 2, 3, and 4 units each, and a 9-measure coda, whose measures are grouped according to the number sequence 2–3–4 (see Figure 37). According to Cage,[113] the first section is the exposition; the next four sections represent the development. The division of 16 units into five sections is based on the number sequence 4–3–2–3–4, whose influence was established even inside the units: every unit is made up of 16 measures and is divided into five phrases according to 4, 3, 2, 3, and 4 measures (as can be seen in Figure 38).

The analysis reveals the totality of the idea of "duration blocks", because in the same relationship of time proportions the composer sought to link a variety of parameters and repeat on a microlevel (measure groups of individual parts) the composition's macro structure.

113 Based on the composer's annotation at the beginning of his score (John Cage: *First Construction*, Edition Peters, 1962).

Figure 37. Cage, First Construction, *Coda. Organization of measures according to numbers 2–3–4*

Figure 38. Cage, First Construction. *Fragment of the first unit, mm. 1–11*

1.1.2. Fibonacci and Number Sequences Derived from it

In today's practice of musical composition an especially active and conscious organization of sound space can be witnessed in the Fibonacci numbers. According to Valeria Cenova, the attention of composers is drawn to this phenomenon by the opportunity to create asymmetry and irregularity in their music and to break mechanical proportions, because such a sequence allows music to "breathe" (Cenova 2000: 51).

Already in the beginning of the 20th century, composers accepted the Fibonacci sequence and the phenomenon of the Golden Ratio as a composition's standard of perfection. To paraphrase Emil Rozenov, even stylistically different musical compositions share the same quality – the manifestation of the Golden Ratio, which controls the music material as an expression of natural beauty (Rozenov 1982: 120–1). Ernő Lendvai, a well-known researcher of Béla Bartók (1881–1945), also believed that the Golden Ratio is probably the most important aspect of a musical composition's architectonics and that it influences all of a composition's parameters. He proved these principles based on musical examples, having researched the relationship between tones and the peculiarities of the rhythmic process.

Further, several examples of creative practices are analyzed on the basis of how the Golden Ratio and Fibonacci numbers influence a musical composition's structure and if the choice of climax (culmination) is made, whether it dictates choice of tone pitches, musical rhythm, grouping of measures, or time signature. The impact of the Golden Ratio can be identified in the proportions between a musical composition's parts as well. For example, in Bartók's *Music for Strings, Percussion and Celesta* (1936) the Golden Ratio determines the micro- and macrostructure levels: the division of the introduction into smaller sections (a microstructural design) is based on the ratio of universal beauty (0.618), the relationship of the length of the first section and the entire composition in eighth notes matches the ratio (3974 : 6432 = 0.6178482…), etc.

Diane Luchese states that another Hungarian composer, whose early work was influenced by Bartók's music, György Ligeti (1923–2006), in his piece for organ *Volumina* (1961–2), applied the logic of the Golden Ratio to the duration of sections with different types of clusters (more see Luchese 1988). According to Ligeti himself, certain Fibonacci numbers are important to the structure of the first section of his *Apparitions* for orchestra (1958–9): in m. 71, at the 144th quarter note, as the bass plays *tremolo*, the first part is divided into two sections; in the second section a striking change in timbre occurs analogously in this sections' 55th quarter note – here an uninterrupted cluster comes in, which plays until the very end of the first part.

The initial Fibonacci numbers, 1, 2, 3, 5, 8, and 13, also dominate the structure of Spanish composer Cristóbal Halffter's (born 1930) *Fibonaciana*, Concert for flute and orchestra (1969). Having analyzed Luigi Nono's cantata *Il canto sospeso* (1955–1956), the durations of the twelve-tone scale of the second part can be notated symmetrically according to the Fibonacci sequence 1–2–3–5–8–13–13–8–5–3–2–1 (Cenova 2000: 47).

How Fibonacci numbers can influence a composition's rhythmic design is illustrated by Sofia Gubaidulina's (born 1931) ensemble *В начале был ритм* for percussion (In the Beginning there was Rhythm, 1984): the numbers 1, 2, 3, 5, 8 pop out in the rhythm of the kettledrum solo. In Gubaidulina's twelve-part composition *Слышу... Умолкло...* for symphony orchestra (I hear... Silence..., 1986), in the odd-numbered parts I, III, V and VII, Fibonacci numbers dictate the time signatures accordingly: 144/4, 89/4, 55/4 and 34/4; in the IXth part, as the composition's silent culmination – in the mute space the conductor's hands follow the rhythm according to yhe Fibonacci sequence (each of the gestures is schemed in detail by the composer).

In Balakauskas' Symphony No. 2 (1979) a slightly transformed sequence with Fibonacci first numbers 1, 2, 3, 5, 8, and 14 (the latter number does not belong in the sequence) influences the segmentation of tone scale and the organization of the progressive rhythm (Daunoravičienė 2000: 92–3).

Besides the Fibonacci numbers, other examples of the sequence are noteworthy in contemporary musical composition practice. For example, analyzing Gubaidulina's work and sketches, Valeria Cenova (Cenova 2000) posits that the Russian composer's musical manipulations are performed based on Édouard Lucas' series and the *Série Évangélique*. According to Johnson, the tonal structure in his composition *Narayana's Cows* in three voices for an undefined ensemble and narrator (1989) was adjusted by a number series, which also inspired the composition's title – the 14th century Indian mathematician Narayana Pandit's (1340–1400) mathematical formula (Johnson *Editions 75*):[114]

1–2–3–4–6–9–13–19–28–41–60 and so on

114 Narayana's numeric formula is similar to the Fibonacci number sequences, but here the two final members are not summed, but the last and the third from the last $S_N = S_{N-1} + S_{N-3}$. In this number sequence, one can see the proportions of the relationship of harmony: every second number in the sequence is larger than the preceding number ca. 46 percent. To be precise, the ratio $\frac{S_{N+1}}{S_N}$ approaches the largest root of the equation $x^3 = x^2 + 1$, which is approx. 1.4655.

Derek Bourgeois. Symphony for Organ, Op. 48 (1975)

The especially prolific English composer Derek Bourgeois (born 1941) has written over 100 symphonies. He dedicated the third part of his Symphony for Organ (1975) to Fibonacci and titled it *Passacaglia di Fibonacci*. Having completed an analysis of the score, it became apparent that the various parameters of the composition were influenced by the Fibonacci numerical algorithm. For example, the entire third part is made up of 144 measures that use 13 different time signatures and five tempo markings; *Largo maestoso* lasts 8 measures, then *l'isteso tempo* – 5 measures, and so on. The time signatures are also written using Fibonacci numbers: 1/8, 2/8, 3/8, 5/8, 8/8, 13/8, etc.

The 24-tone subject is worth its own discussion. The subject is continually repeated in the bass part and composed with the technique of passacaglia (*basso ostinato*); its length in measures is 13, and the number sequence expresses the arrangement of the melody-line in semitones as follows:

1–1–2–3–7–8–1–3–2–7–7–0–5–7–2–3–1–4–7–9–10–11–11

An internal logic of the 24-tone subject at first is hardly noticeable and sounds even chaotic. But in fact the subject melody is characterized by the relationship between tones written in true and "secret" Fibonacci numbers. At first glance, some members of the Fibonacci sequence are hardly noticeable. That is because the range in semitones of a few intervals does not match the Fibonacci numbers – that is the major third (M3, 4 semitones), the fifth (P5, 7 semitones), the major sixth (M6, 9 semitones), and the minor and major sevenths (m7, 10 semitones, & M7, 11 semitones). Their range in semitones is equal to the numbers 4, 7, 9, 10, and 11. However, the inversion of these music intervals matches the Fibonacci numbers: a M3 inversion is a minor sixth, whose range is 8 semitones; correspondingly, a P5 inversion is a perfect fourth and the range in semitones is 5; M6 inversion is minor third (3 semitones); m7 inversion is major second (M2) and 2 semitones; M7 inversion is a minor second (m2) and 1 semitone. Moreover, the numerical sequence, rewritten in Fibonacci numbers, is of a mirror symmetry origin, and its center is marked by the only perfect unison (P1) with the numerical expression 0 (see Figure 39).

Narayana's system is based on the laws of nature – the process of a cow bringing a calf into the world. It was stated that from every one cow producing one calf a year, who after four years also produces a calf, a number sequence emerges:

1	2	3	4	6	9	13	19	28	41	60…
				(1 + 3)	(2 + 4)	(3 + 6)	(4 + 9)	(6 + 13)	(9 + 19)…	

This observation allows one to surmise that the composer did not seek to use Fibonacci numbers in an obvious manner, but rather in a creative one. A similar incentive to creativity can be explained by the impact of the Fibonacci numbers on the time parameter: the number of measures are not recorded as usual (every 5 or 10 measures), but in those parts of the score that correlate with Fibonacci numbers. Bourgeois had an even more clever idea by trying to hide Fibonacci numbers, by looking at the logic of sections III, IV, and V. Their volume in measures (respectively 65, 24, and 10) does not correspond to the Fibonacci numbers. However, if one were to remove the number 10 (extent of section V) from the number of measures of section III (65) or if one were to add the number 10 to the section IV (24), we would get Fibonacci numbers 55 and 34 (see Figure 40).

Figure 39. Bourgeois, Symphony for Organ, Op. 48, part 3 Passacaglia di Fibonacci. *Arrangement of musical intervals according to Fibonacci numbers, mm. 14–28*

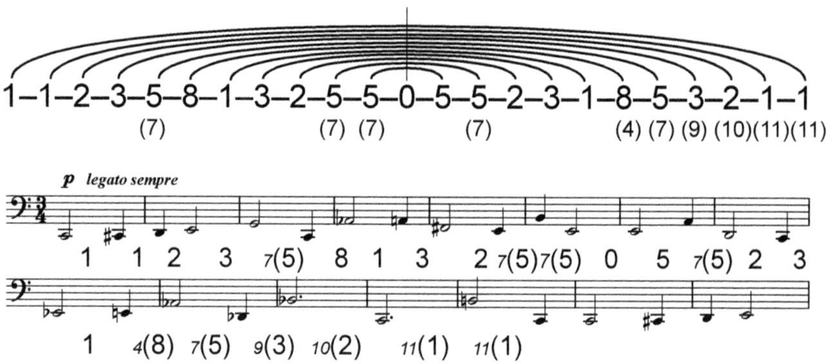

Figure 40. Bourgeois, Symphony for Organ, Op. 48, part 3 Passacaglia di Fibonacci. *Arrangement of time signatures and number of measures of sections according to Fibonacci numbers*

Sections	1	2	3	4	5	6	7	8	9	10	11	12	13	14	15	16	17	18	19	20	
Meter signatures	8/8	5/8	3/4	5/4	8/4	13/4	3/4	21/4	13/4	8/4	5/4	3/4	2/4	1/4	1/8	2/8	3/8	5/8	8/8	13/8	
Number of bars	8	5	*65*	*24*	*10*	3	13	3	2	1	1	1	1	1	1	1	1	1	1	1	Total 144
			–	+	↑																
			10	*10* ←																	
			55	**34**																	

Steve Reich. *Clapping Music* (1972)

I would like to argue that we see an example of the influence of the constructive Fibonacci sequence in Reich's minimalist piece *Clapping Music* (1972). While analyzing this composition, it became apparent that the logic of this compact composition follows the Fibonacci numbers, arithmetic progression and the laws of symmetry.

Let us look at the basic rhythmic pattern ♪♪♪ 𝄾 ♪♪ 𝄾 ♪ 𝄾 ♪♪ 𝄾 where the first performer claps consistently. This rhythmic structure can be reduced to a formula of arithmetic progression. The tones, rests, and total number of all elements are in the relationship of 4 : 8 : 12, because 4 is the sum of eighth note rests, 8 is the sum of eighth notes, and 12 is the total sum of musical elements; accordingly 4 : 8 : 12 → 1 : 2 : 3.

The second performer's score was composed of a rhythmic pattern and 11 of its cyclic permutations. The permutations were created consistently by carrying the first member to the end. Here one can see the manifestation of mirror symmetry. The symmetry was established in the number sequence, which was written down after having counted the number of clapping tones of the second performer from one rest to the next (see Figure 41).

The Fibonacci numbers appear in several ways in this piece. The sum of the number of letters in the composition's title *Clapping Music* is 13. Each word is made up of 8 or 5 letters. The rhythmic invariant of the piece is made up of 8 eighth notes. The rests divide the row of eighth notes into four sections according to the following: 3 eighth notes, 2 eighth notes, 1 eighth note and 2 eighth notes. Most likely, it is not a coincidence that the Fibonacci numbers influence two time signatures that were written in at the beginning of the piece ♩ = 144 and ♩ = 168 (the latter number is not a member of the Fibonacci sequence, but is equal to the multiplication of F_6 and F_8: 8 x 21). The full number of measures in this piece is the number 13 (see Figure 42).

Figure 41. Reich, Clapping Music. *Manifestation of mirror symmetry in the part of the 2nd performer (the numbers indicate the quantity of clapping tones between the rests)*

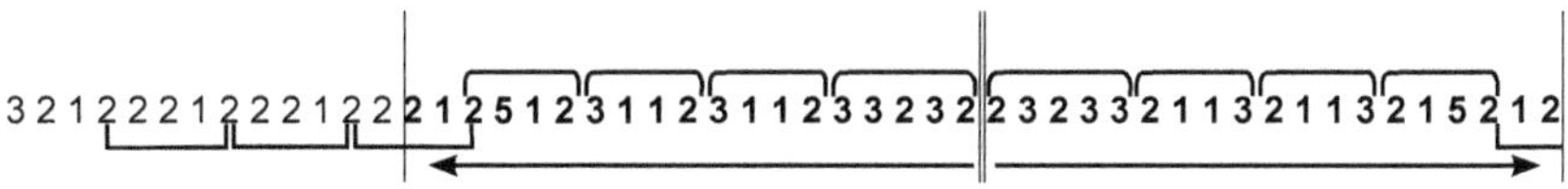

Figure 42. Reich, Clapping Music. *Structural diagram*

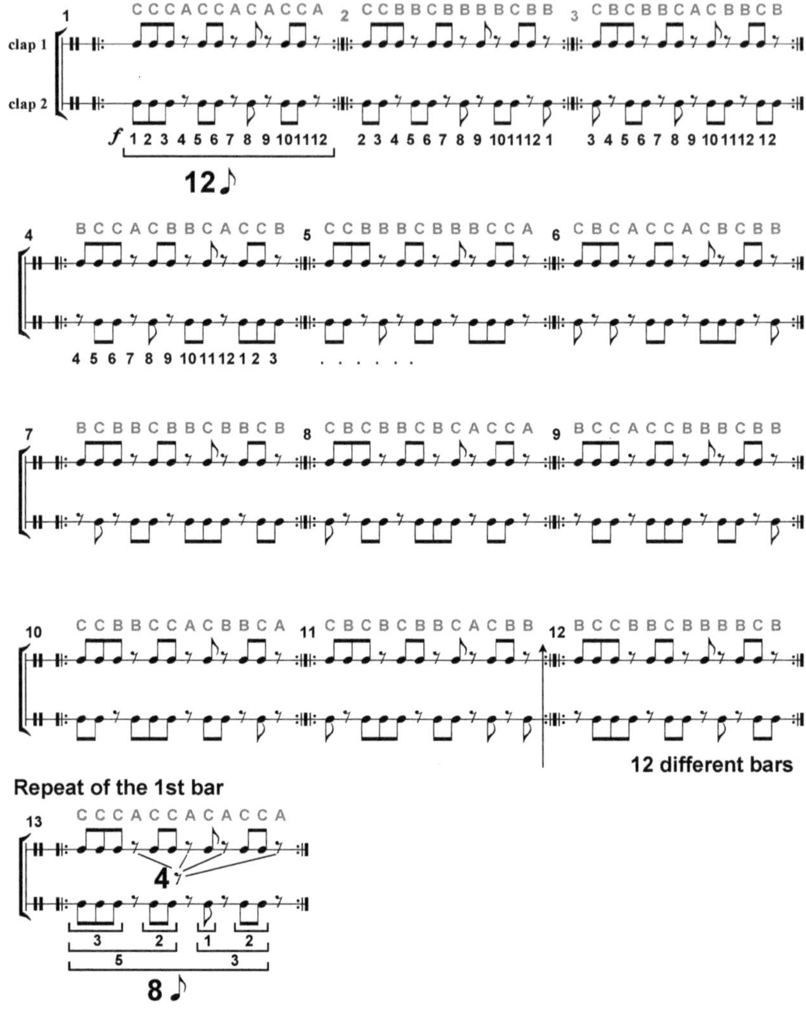

Karlheinz Stockhausen. *Klavierstück IX* (1961)

The inheritance of the Fibonacci numbers in terms of time signatures and tonal groups is apparent in the work of Karlheinz Stockhausen (1928–2007), which have already achieved the status of becoming anthologized. Strong examples of the use of Fibonacci numbers can be found in the following original compositions from the 1960s: *Klavierstück IX*, *Plus-Minus*, *Mikrophonie I*, *Zyklus*, *Stop*, *Adieu* or *Telemusik*. For example, in the notes to the score of *Zyklus* (The Cycle, 1959) the composer explains that he regulated the proportions between tonal groups according to the numbers 1, 2, 3, 5, and 8. Creating the structural scheme of the wind quintet *Adieu* (1966), Jonathan Kramer proved that in measures the entirety of the composition corresponds to the relationship of the Fibonacci numbers (Kramer 1973: 127).

The influential use of the Fibonacci numbers appears in the organization of time signatures in *Klavierstück IX* (Piano Piece No. 9, 1961). Here, it would seem the entire composition is "dissected" by the Fibonacci numbers. We see this in the record of time signatures in eighths only (Fibonacci number 8). Additionally, the composer wrote down the numbers of this progression, indicating the number of beats in the bar. It is apparent already in the change of time signatures in the first 13 measures:

13/8–2/8–21/8–8/8–1/8–3/8–8/8–1/8–5/8–13/8–2/8–5/8–3/8

Having analyzed the composition's coda, it became obvious that not only the meter, but the number of tones in each measure, in addition to how they were organized in parts for the right hand and for the left hand, correspond with Fibonacci numbers (see Figure 43). A few numbers in the coda (for example, 6, 10, and 11) are not mentioned as part of the progression. However, it should be considered that Stockhausen also used derivatives of the Fibonacci sequence. According to Gregg Wager, the composer applied the sum of three or more adjacent numbers (Wager 1998: 94). In this manner, he arrived at the following sequences:

6–11–19–32–53–87–142 …
or
6–10–16–42–68–110 …

These sequences are created by adding all previous Fibonacci numbers or the last three numbers:

$1 + 2 + 3 = \mathbf{6}$ $\quad$ $1 + 2 + 3 = \mathbf{6}$

$1 + 2 + 3 + 5 = \mathbf{11}$ $\quad$ $2 + 3 + 5 = \mathbf{10}$

$1 + 2 + 3 + 5 + 8 = \mathbf{19}$ $\quad$ $3 + 5 + 8 = \mathbf{16}$

$1 + 2 + 3 + 5 + 8 + 13 = \mathbf{32}$ and so on[115] $\quad$ $5 + 8 + 13 = \mathbf{26}$ and so on

Members of the first number sequence 6–11–19–32–53–87 … were determined as forming the coda. The coda consists of 8 groups. By counting their duration in lengths of eighth notes, the first group is equal to 1 eighth note, the second is equal to 3 eighth notes, the third is equal to 6 eighth notes, the fourth is equal to 11 eighth notes, the fifth is equal to 19 eighth notes, the sixth is equal to 32 eighth notes, the seventh is equal to 53 eighth notes, and the eighth is equal to 87 eighth notes.

Figure 43. Stockhausen, Klavierstück IX, *Coda. Record of meter signatures and selection of numbers of tones in each measure according to Fibonacci numbers*

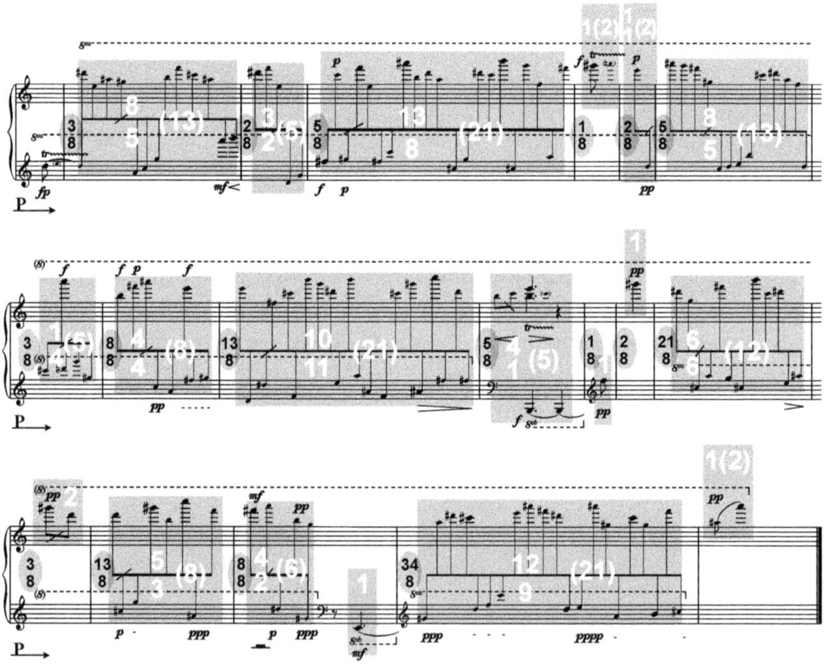

115 According to Kramer (Kramer 1973: 124), the number sequence 6–11–19–32–53–87–142… is derived from the Fibonacci sequence by applying the formula $F_n - 2$: (3–2 = 1, 5–2 = 3) 8–2 = **6**, 13–2 = **11**, 21–2 = **19** and so on

1.1.3. The Prime Numbers and other Mathematically Determined Sequences

In the practice of composing contemporary music, in addition to the traditionally defined proportions of Antiquity, the Fibonacci sequences, and principles of other derivatives, composers used other mathematically determined series or themselves created number sequences. The following examples illustrate this practice.

Alain Louvier. *L'Isola dei Numeri* (1991)

The French composer Alain Louvier (born 1945) often derives inspirations for his compositions from various mathematical phenomena. His collection of 19 pieces for piano, *Agrexandrins* (1981–92), was composed using complicated progressions. The inspiration and algorithm for his piece *Eclipse* (1999) came from the geometric trajectory of earth's elliptical orbit around the sun. A topographical map was woven into the material of the music in his composition *Quatre paysages* (1999). Unique properties of the Pascal triangle were transferred into relationships of musical tones in his piece *Triangle* (1997). Louvier's *Hommage à Gauss* for violin and orchestra (1968) was composed in honor of the mathematician Carl Friedrich Gauss, who was renowned for his phrases "mathematics is the queen of the sciences"[116] and "God does arithmetic."[117]

This composer's work shows how prime numbers can be applied to music as well. In his annotations to the collection of 6 etudes for piano, *L'Isola dei Numeri* (The Island of Numbers, 1991, 3 notebooks of 2 etudes each), Louvier comments on the inspiration for his piece:

> It is an ever-changing archipelago from island of cliffs, from tones, from a few shores of toccatas, from isolated shores, from which the horizon opens up in an endless chain of numbers.[118]

---

116 German "Die Mathematik hielt Gauss um seine eigenen Worte zu gebrauchen, für die Königin der Wissenschaften [...]", cited in Wolfgang Sartorius von Waltershausen: *Gauss zum Gedächtniss*, Leipzig: S. Hirzel, 1856, p. 79.

117 Attributed to Gauss, quoted in Alan L. Mackay: *A Dictionary of Scientific Quotations*, Bristol and Philadelphia: Institute of Physics Publishing, 1991, p. 100.

118 Citation according to Alain Louvier: *L'Isola dei Numeri*, Paris: Alphonce Leduc, 1992, p. I.

This idea was realized by choosing a sequence of prime numbers and the indications of the intervals between them: 1–2–2–4–2–4 … This number row is illustrated in the following figure:

2 – 3 – 5 – 7 – 11 – 13 – 17 – 19 – 23 – 29 – 31 – 37 – 41 – 43 – 47 – 53 – 59 – 61 – 67 – 71 – 73 and so on

1 2 2 4 2 4 2 4 6 2 6 4 2 4 6 6 2 6 4 2

According to the composer, in the fifth etude *Toccata serpentin,* the pitch of 11 tones is determined by the differences between the prime numbers. Incidentally, Louvier's decision to use not all 12 chromatic tones, but one less, could be described as a conscious intention to relate the tonal sequence with the prime number 11. It was established that the tonal pitches were appropriately matched with these differences between prime numbers:

tone *c* – 1
c-sharp/d-flat – 2
d – 4
d-sharp/e-flat – 6
e – 8
f – 10
f-sharp/g-flat – 12
g – 14 and so on.

By taking into consideration Louvier's hint that he counted backwards, the arithmetic differences at the beginning of the sequence (1, 2, 2, 4, 2, 4, 2, 4, 6 and so on) can be deciphered from the conclusion of the etude to reveal the scale (see Figure 44).

It was noticed that in the etude's time signature symbols, the prime numbers 1 through 37 were used. The time signature for eighth notes was written down as an expression of only 1/8, while the time signature of a quarter note has seven different variations, 1/4, 2/4, 3/4, 5/4, 7/4, 11/4, and 13/4 (1 is an invertible element in the ring of integers, and 2, 3, 5, 7, 11, and 13 are prime numbers). Meanwhile, there are even 11 different time signatures of a sixteenth note. In those all the prime numbers from 1 to 37 are applied (1/16, 2/16, 3/16, 5/16 … 37/16).

Figure 44. Louvier. L'Isola dei Numeri. *Above: general tone scale; below: end of Etude No. 5* Toccata serpentin, *organization of tone scale intervals and meter signatures according to prime numbers*

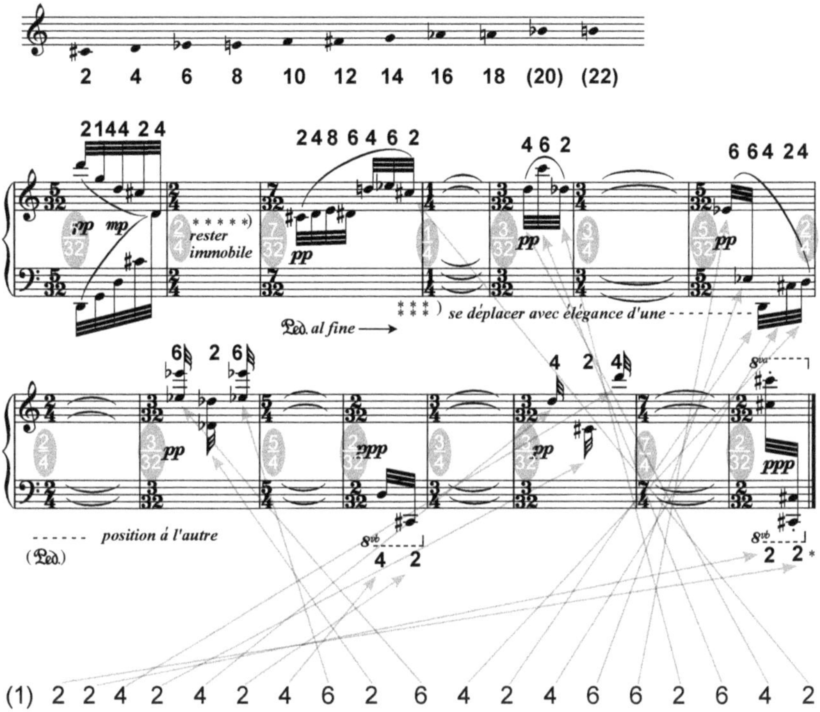

Tom Johnson. *Music for 88* (1988)

It could be said that one of the goals of Johnson's nine-piece cycle *Music for 88* for piano (1988) was to familiarize the performer and the listener with particular mathematical formulas. It were as though in his music he wanted to recreate mathematical functions: squaring numbers or calculating the multiplication table. He structured the tonal sequences according to Mersenne numbers or carried the form of the geometric triangle over into the score and so on.[119]

119 This is witnessed by the titles of the nine pieces. For example, in the piece *Eighty-Eights (88)* all the piano keys are presented. The piece *Mersenne Numbers* is composed based on the mathematician Marin Mersenne's number sequence. In the third piece of the cycle, *Multiplication Table*, the multiplication table is recorded. In the piece

An analysis of the final piece in the cycle *Eratosthenes Sieve* illustrates how tonal pitches are organized according to sequences of prime numbers. The piano keyboard consists of 88 keys. In several stages Johnson equates them to a natural number sequence. In the score only those tones that matched a prime number were written down. These were arrived at by counting the distance between semitones from the lowest key on the piano. Because of the limited range of the piano (88 keys) the numbers used in the first section of the piece were selected from 1 to 88, and accompanied with the consistently repeating motif *c-sharp – d-sharp – g – a* and its variants (see Figure 45). The composer applied this principle for a second time with a number sequence from 89 to 176. In total, he repeated this action 23 times (23 is a prime number), until he arrived at the prime number 1987. At this point, Johnson stopped, as though wanting to commemorate the date of composition (1988).

It is obvious that Johnson basically counted semitones in order to establish equivalents of prime numbers on the keyboard. However, the composer had an even more interesting idea regarding the rests. It was established that the duration and changing of rests (one eighth note rest, one quarter note rest, two quarter note rest, and so on) was determined by the amount of odd numbers between the pairs of consecutive prime numbers (one odd number was identified with an eighth note rest, two with a quarter-note rest, and so on). For example, the equivalents of the prime numbers 7 and 11, the tones $d\text{-}sharp_1$ and g_1, were separated by one eighth-note rest (that is between the numbers 7 and 11 there is one odd number 9). The equivalents of the numbers 89 and 97, the tones a_2 and f_1 were separated by the rest of three eighth notes (between the numbers 89 and 97 there are three odd numbers, 91, 93, and 95; see Figure 46).

Squares those actions that are performed by calculating the square of a number are interpreted by piano keys. In the composition *Triangles* we hear an interpretation of the geometric form of the triangle. In the composition *Abundant Numbers* the theory of abundant numbers is expressed in musical tones. In the piece *Euler's Harmonies* Leonhard Euler's (1707–1783) mathematical harmony theory is presented. In the piece *Pascal's Triangle* the 17th century mathematician and philosopher Blaise Pascal's (1623–1662) triangle is expressed through musical tones. In the piece *Eratosthenes Sieve* musical tones are used to paraphrase the algorithm of discovering prime numbers by the famous Ancient Greek mathematician, astronomer, and geographer Eratosthenes (Ερατοσθένη, 276–194 B.C.).

Figure 45. Johnson, Eratosthenes Sieve, *beginning. Equivalents of prime numbers and tones*

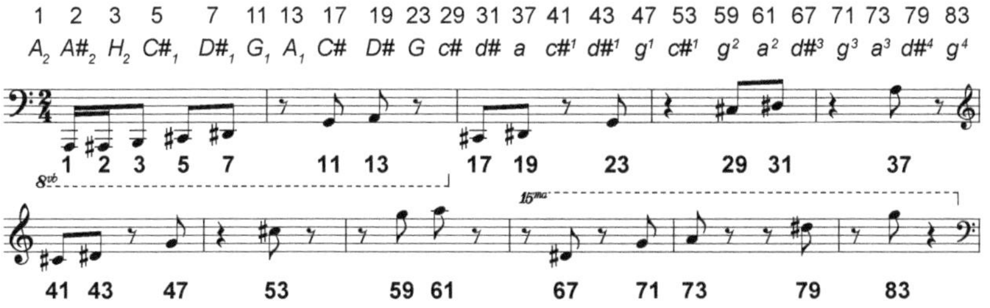

Figure 46. Johnson, Eratosthenes Sieve, *second section. Organization of rests according to odd numbers*

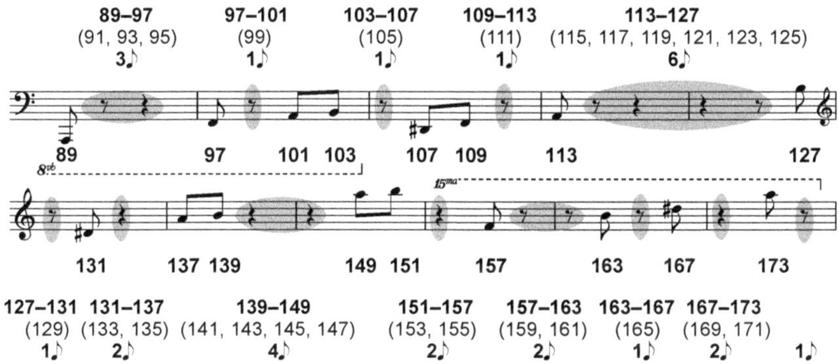

An elementary composing process is typical of the second piece *Mersenne Numbers*, which in honor of the 17th century French mathematician, Mersenne, recreates principles of number sequencing. According to the composer, Mersenne caught his attention because his birth date, 1588, is associated with the number 88, the number of keys on a piano (another connection is that Johnson's cycle was composed in 1988). Johnson structured the tones in this piece by merely comparing Mersenne numbers with the number of keystroke in the right hand (see Figure 47). This way only the nine numbers 1, 3, 7, 15, 31, 63, 127, 255, and 511 were involved. Later, the composer turned round the number sequence by giving the piece mirror symmetry.

These elementary operations with numbers were supplemented with an intriguing mathematical construction of tone scale. To define the source of the scale it was necessary not to rely on traditional counting. By analyzing the regular "offspring" of the tone scale, it was established that it integrated a repetitive cumulative counting process. It is an algorithm that is used to count how vegetation grows, which is called L-system (Lindenmayer system). Typical to this system is the recursive perpetual growth.[120] The use of this system in music creates a composition with a self-similar tonal structure, which is often found in Johnson's work. In *Mersenne Numbers*, the technique of repetitive cumulative counting is identified as separate tones replaced by integers, as follows: *c* – 1, *d-sharp* – 2, *e* – 3, *g* – 4, *a-flat* – 6, *b* – 7. By writing out the combinations of the first six tones, a certain interchange of Mersenne numbers can be established.[121]

By applying the example of the L-system, it became clear that in Johnson's work in order to create sound from Mersenne numbers, a consistent action of mathematical counting was used. According to this action, each former number (tone) in the sequence in the new line was changed into a combination of a few numbers (tones), in order to match to a certain changing step:

- the first element, an axiom, remains unchanged; in other words, *c* changes to *c* or 1 → 1;
- the second member is changed by a ternary combination; i.e., the tone *d-sharp* is changed to the three-tone motif *d-sharp – e – d-sharp* or 2 → 232;
- the third member is changed by the appropriate ternary combination; i.e., the tone *e* is changed to *e – g – e* or 3 → 343.

In this manner, from the combination of the first three numbers, 1–2 – 3, its first variation was created, 1–232 – 343. This group of seven numbers is equated to Mersenne number 7 and its musical motif:

c – d-sharp – e – d-sharp – e – g – e

120 Recursion, or iteration – the return of former elements and their combinations.

121 The process of the behavior of plants, as defined by the L-system, can be seen in other examples in compositions of contemporary music as well. This kind of music composing is associated with the manifestation of innovative mathematical tendencies and can be found in the section "L-system Formalities in Music" in this book, pp. 213–218.

However, by further assigning a sound to a number, such as 15, 31, 63, and to the remaining Mersenne numbers, Johnson cheats on the mathematical precision of calculation. Not all the numbers are changed by a ternary combination. This is because in the action of the analogically repetitive accumulation with the derived number sequence 1–232 – 343, in total in the new sequence we get 19 members, and not the desired Mersenne number 15. Therefore, trying to create a Mersenne number sequence, the composer applied the conversion to a ternary combination every second step as follows:

- two of the first numbers are changed by a ternary combination,
- the two numbers that come after the first two numbers remain unchanged,
- the next two members are changed by a ternary combination and so on (see Figures 48 & 49).

Figure 47. Johnson, Mersenne Numbers, *beginning. Diagram of mirror symmetry and the arrangement of keystroke according to Mersenne numbers*

1	2	3
1	2–3–2	3–4–3
1	232–343–2	3–454–343

Figure 48. Johnson, Mersenne Numbers. *Arrangement of tone scale according to Mersenne numbers*

1 c^1
3 $c^1-d\#^1-e^1$
7 $c^1-d\#^1-e^1-g^1$
15 $c^1-d\#^1-e^1-g^1-a\flat^1$
31 $c^1-d\#^1-e^1-g^1-a\flat^1-b^1$
63 $c^1-d\#^1-e^1-g^1-a\flat^1-b^1-c^2$
127 $c^1-d\#^1-e^1-g^1-a\flat^1-b^1-c^2-d\#^2$
255 $c^1-d\#^1-e^1-g^1-a\flat^1-b^1-c^2-d\#^2-e^2$
511 $c^1-d\#^1-e^1-g^1-a\flat^1-b^1-c^2-d\#^2-e^2-g^2$

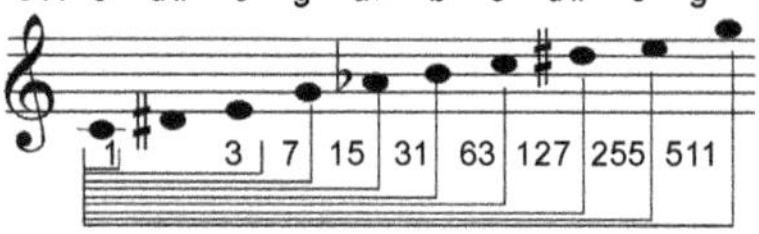

1	c^1	1
3	$c^1-d\#^1-e^1$	1 2 3
7	$c^1-d\#^1-e^1-d\#^1-e^1-g^1-e^1$	1 232 343
15	$c^1-d\#^1-e^1-d\#^1-e^1-g^1-e^1-d\#^1-e^1$…	1 232 343 23 454 343
31	$c^1-d\#^1-e^1-d\#^1-e^1-g^1-e^1-d\#^1-e^1$…	1 232 343 23 454 343 23 454 565 43 454 343
63	$c^1-d\#^1-e^1-d\#^1-e^1-g^1-e^1-d\#^1-e^1$…	1 232 343 23 454 343 23 454 565 43 454 343 23 454…

Figure 49. Johnson, Mersenne Numbers. *Arrangement of the melody according to L-system algorithm*

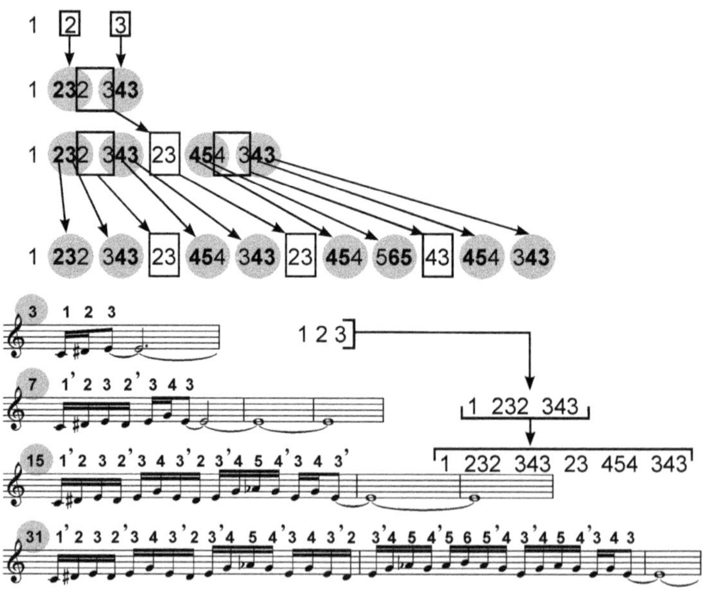

1.2. The Renewal of Polytempo, Polyrhythm, and Polymeter

Metric consonance and dissonance, manifestation of combinations of different time signatures, or the conflict between meter and rhythm in contemporary music can be described as the reconstruction of the practice of using polytempo, polyrhythm, and polymeters in Renaissance music. By using these practices, contemporary composers are able to produce refined sound effects, when creating a work's dynamics, and the polyphony of layers of different motions.

The score of Stravinsky's *The Rite of Spring* (1913) from the early 20th century is often pointed out as a widely anthologized example. According to Eliot Woodruff, by analyzing this composition's rhythmic features, in the background[122] there is a typical unchanging meter, despite the fact that in the foreground[123] various complicated metric values change irregularly (more see Woodruff 2006). For example, a "chaotic" shift in time signatures is detected in the part "Ritual of Abduction", mm. 1–31:

time signature	9/8	4/8	5/8	9/8	12/8	9/8 (4/8+5/8)	(5/8+4/8)	6/8	7/8	3/4 …
mm.	1–8	9	10	11–8	19	20	21	22–9	30	31
number of mm.	8	1	1	8	1	1	1	8	1	1

A polymetric phenomenon, when the marked time signature does not match the sound motion that is heard, can also be found in musical compositions of the Romantic era. For example, Beethoven, in the beginning to his finale part of the Trio, Op. 1 No. 1, seemingly ignored the bar lines and transferred the first (stressed) beat to the upbeat. The metric conflict in Schumann's *Scherzo* in F minor, Op. 14 was influenced by the rhythmic accents not matching with the bar lines. In Brahms' compositions, it was also typical for the composer to deny the limits of bar lines. For example, the use of slurs of the melodic phrases in *Intermezzo*, Op. 118 No. 2, mm. 38–43, shows a fusion of duple time and triple time, although the composition is written in a 3/4 time signature (see Figure 50).

In 20th–21st century music, this phenomenon with its mathematical basis reminds one of the different coordination of modus in *Ars Nova* practice, the techniques of the Flemish School composers during the Renaissance, which were noteworthy for the rational combinations of mensuras and techniques of proportional canons. In contemporary Lithuanian music, the composer Rytis Mažulis (born 1961)

122 German *Hintergrund*, according to Heinrich Schenker's *Ursatz* theory.

123 German *Vordergrund*.

Figure 50. Brahms, Intermezzo, *Op. 118 No. 2. Exchange of duple time and triple time*

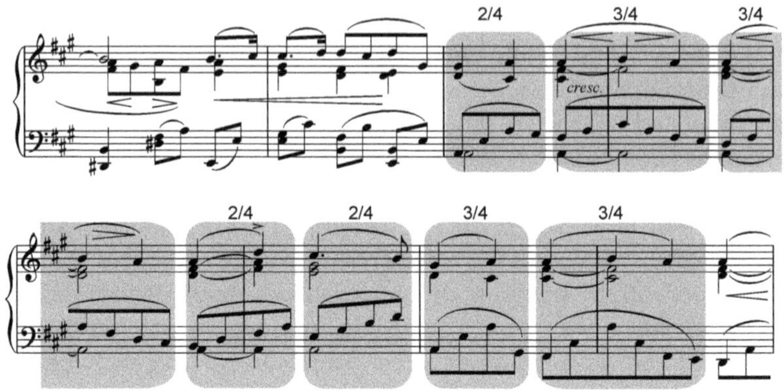

especially widely applies the Renaissance mensural phenomenon by creating several metric layers into his composition's sound space. His scores *Palindrome* (1996), *Cum essem parvulus* (2001) and *ajapajapam* (2002) are based on polytempo relationships and are composed like palindromic polytempo canons. Ričardas Kabelis (born 1957) in his *Oriono diržas* for computer (Orion's Belt, 1996) used complicated mathematical ratios between different tempos, constructing a "microrhythmic polytempo structure" (Mažulis 2001: 69). These Lithuanian composers are known for using a composing technique that is similar to the creative work of Ligeti or Nancarrow. For example, the Ligeti etudes for piano or Nancarrow's especially complicated studies for player piano are typified by simultaneously using different tempos, series of tempos, or proportions. The musicologist Hannes Schütz describes this phenomenon as a resonance of the 14th century *Ars subtilior* (see Schütz 1997).

John Adams. *Short Ride in a Fast Machine* (1986)

The effects of metric irregularities is often expressed in compositions of minimalist music. Typical of this style is the constant repetition of certain melodic-rhythmic figures, which cross beyond the framework of the metric notations, as though ignoring the bar lines.

One example of metric conflict can be seen in John Adams' (born 1947) score for orchestra *Short Ride in a Fast Machine* (1986). At the beginning of the piece the time signature is 3/2; the bar lines strictly follow the three-two time. However, this order is destroyed by trumpet accents, offering an alternative metric pulsation, for example 2/2, 3/4, 5/4. The conflict between the marked actual time signature and the motion that is heard by the listener already happens on the first page of the score (see Figure 51).

Figure 51. Adams, Short Ride in a Fast Machine. *Metric conflict in mm. 1–19*

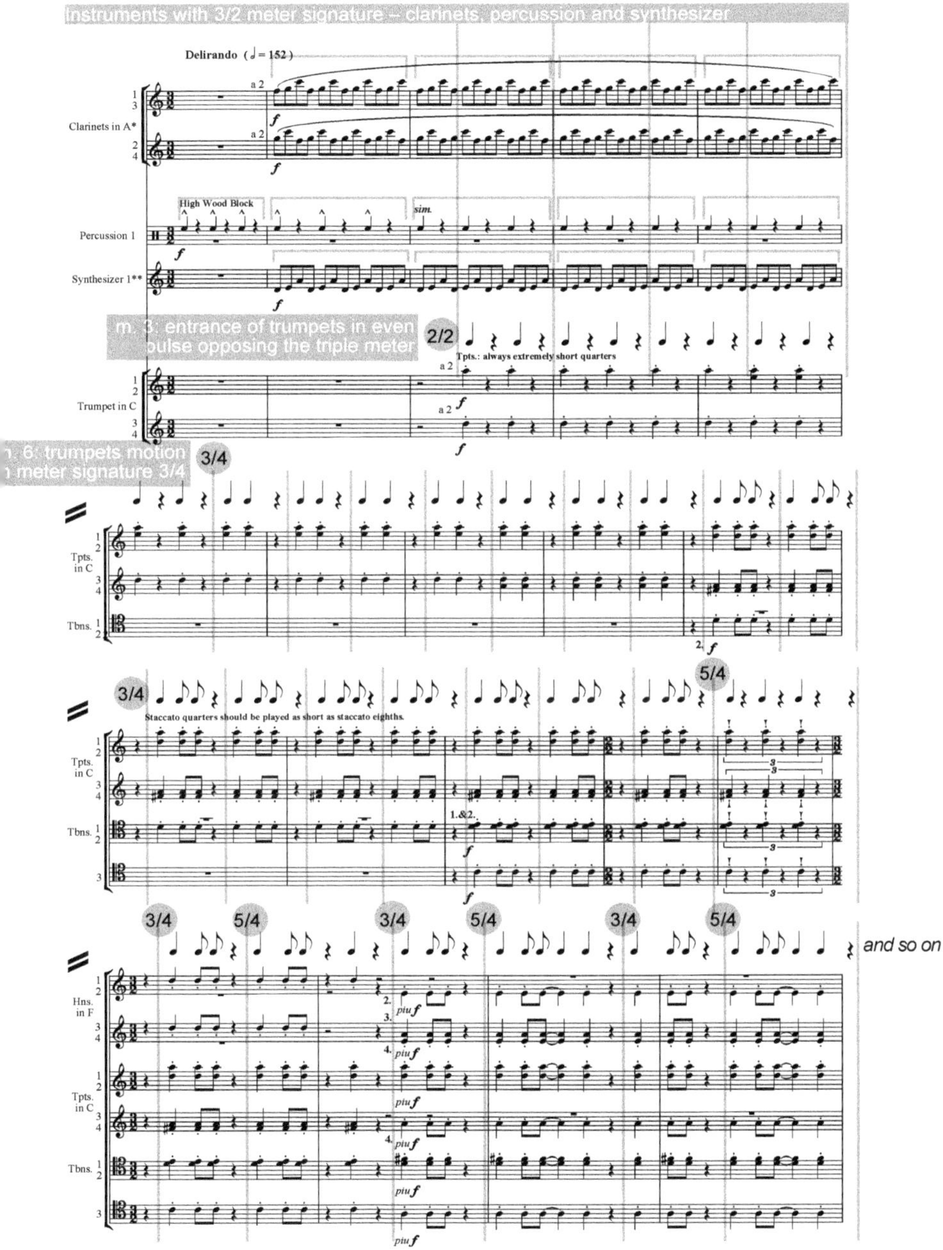

Conlon Nancarrow. Study for Player Piano No. 15 (premiere in 1962)

Conlon Nancarrow's (1912–1997) experimental transcendental pieces – *Studies for Player Piano*[124] – are beyond the capabilities of a live performer and must be performed by man-made mechanical instruments. The unique canon and multitemporal solutions perfectly reflect this composer's original touch. Because of his fascination with the player piano, Nancarrow, according to Nouritza Matossian, is often described as one of the most mysterious and eccentric figures in contemporary music (Contemporary Composers 1992: 677). Almost three quarters of his studies are created using the canon technique. Most of them embody an idea of intriguing rhythms and asynchronous tempos. They are composed based on tempo proportions and mathematical calculations. Henry Cowell's (1897–1965) fundamental study *New Musical Resources* (1930) and his propagated ideas about using polyrhythm, polymeters, and polytempos in music had a huge effect on Nancarrow's inspirations. Kyle Gann argues that Nancarrow delved into Cowell's innovations: the opportunity to divide a large rhythmic block (or measure) into a variety of equal elements and by so doing to create the effect of different tempos playing at the same time (Gann 1995: 42). James Saunders describes Nancarrow's passion for manipulating the effects of polytempo as the concept of distorted time in music, proving that it is possible to compose music not beholden to the typical musical gravitation, not limited by a linear system of rhythm, in fact, by denying it and destroying it. Therefore, the composer first sought to direct the listener's attention to the polyphony of tempo and the pulsating rhythm, leaving the melodic relief and harmony in the second plane (Saunders 1996: 10, 30).

Nancarrow himself has said that "time is the last frontier of music" (Thomas 2000: 107). Therefore, by exploiting the possibilities of musical tempo and by using original means to expose the polyphony of this parameter, he created a flexible and constantly changing musical space. For example, in the twelve-line canon in his Study No. 37 the composer wrote down the following number sequence:

$$150–160\,{}^{5}/_{7}–168\,{}^{3}/_{4}–180–187\,{}^{1}/_{2}–200–210–225–240–250–262\,{}^{1}/_{2}–281\,{}^{1}/_{4}$$

This sequence matches with the mathematical relationships of the just chromatic scale. By the way, this sequence is in Cowell's treatise. According to Gann, for a long time Nancarrow's study walls were covered with Cowell's figures and graphs (Gann 1995: 194).

124 *Studies for Player Piano* were composed from the end of the 1940s to the middle of the 1990s. In total, Nancarrow wrote about 60 studies. Most of them were dedicated to one player piano. Studies No. 39–41, 43 and 44 were for two player pianos, and No. 30 – for prepared player piano.

$281^{1}/_{4}$	15:8	*b*
$262^{1}/_{2}$	7:4	*b-flat*
250	5:3	*a*
240	8:5	*a-flat*
225	3:2	*g*
210	7:5	*f-sharp*
200	4:3	*f*
$187^{1}/_{2}$	5:4	*e*
180	6:5	*e-flat*
$168^{3}/_{4}$	9:8	*d*
$160^{5}/_{7}$	15:14	*c-sharp*
150	1:1	*c*

Nancarrow's experiments with the parameters of time in music had a strong influence on many late 20th century composers. We will discuss his influence on Ligeti, when analyzing Ligeti's etude *Désordre*. The idea of a chromatic scale, and most likely the particular composition of Nancarrow (Study No. 37), influenced a few of Stockhausen's creative ideas. For example, the same numerical relationships in the chromatic scale as an equivalent to 12 tempos can be found in Stockhausen's cycle of 12 melodies, *Tierkreis* (Zodiac, 1975). The graphic figure from Nancarrow's Study No. 37 is almost analogical to Stockhausen's tempo diagrams for his piece *Gruppen* (Groups, 1957), which were published in his 1957 article (Gann 1995: 195). John Cage was inspired by Nancarrow's ideas and dedicated some of his lyrics to him.[125]

---

125 Johns Cage's verse was published in Conlon Nancarrow: *Selected Studies for player piano* (Soundings 4), ed. Peter Garland, Berkeley: Sounding Press, 1977, p. 25:

A Long Letter

the musi**C**
y**O**u make
is**N**'t
Like
any **O**ther:
tha**N**k you.

o**N**ce you
s**A**id
whe**N** you thought of
musi**C**
you **A**lways
thought of you**R** own
neve**R**
Of anyone else's.
that's ho**W** it happens.

The unique attributes of his two-voice Study No. 15 (the first performance took place in 1962) are influenced by the relationships of the numerical formula 3/4, which was written down in the beginning of his composition as *Canon ¾*. These notes serve to explain the choice of metronome marks 165 and 220 for the canon, because 165 : 220 = 3/4. By analyzing the score, it was established that the relationship 3/4 determines the shift and disagreement of bar lines between the two voices. This displacement creates the palindromic two-voice juxtaposition, which opens up in a graphic chart of the polytempo canon. In Figure 52 the numbers were marked, which showed the quantity of eighth notes in one bar. This is because the composer wrote down every measure with a different time signature, although everywhere he maintained the same rhythmic value (eighth note): 8/8, 5/8, 6/8, 5/8, 6/8, 9/8, and so on. It was noted that in his metro-rhythm organization Nancarrow used a number sequence from 3 to 11, omitting only number 10 (the time signatures in the score are written out as 3/8, 4/8, 5/8, 6/8, 7/8, 8/8, 9/8 and 11/8; see Figure 53).

Figure 52. Nancarrow, Study for Player Piano No. 15. Palindromic canon

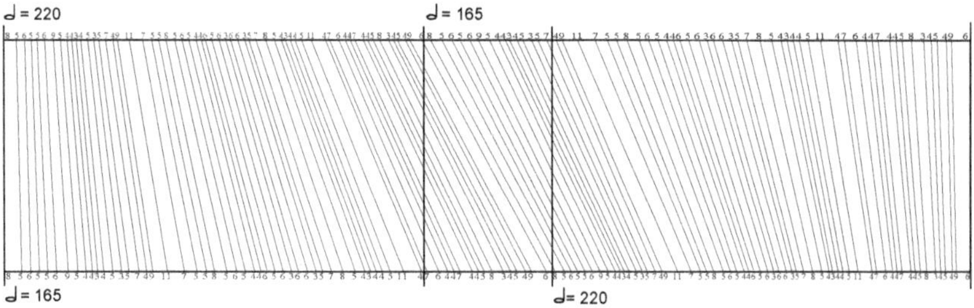

Figure 53. Nancarrow, Study for Player Piano No. 15. Shift and disagreement of bar lines

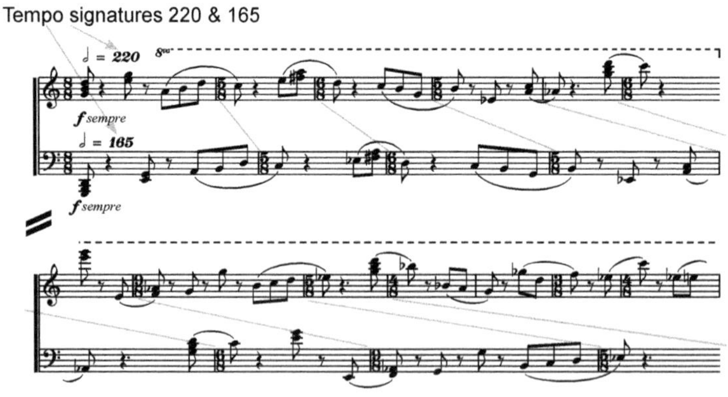

György Ligeti. *Désordre* (1985)

The direct inspiration for Ligeti's etudes for piano is not only this composer's interest in the folk music of Central Africa and his familiarity with the Benoit Mandelbrot's theory of fractals and Heinz-Otto Peitgen and Peter Richter's computer generated fractal images, but also the complicated canons of Nancarrow. He became interested in them in the 1980s. Ligeti was intrigued by the effect of illusionary rhythm. According to the composer himself (Ligeti 1988: 10), the best example would be the third part of his Concerto for piano (1985–8); here the bar line is rendered unnecessary.

Even before Ligeti was influenced by Nancarrow's compositions for the player piano or with African folk music, he had already applied polyrhythmic principles to his piece for two pianos, *Monument* (1976). In this composition, two pianists perform the same phrase in a different meter – duple time and triple time. This is also illustrated in an earlier idea by Ligeti, the effect of which the composer compares with the house of an old widow that is filled with old ticking clocks (Clendinning 1996: 2). This is a composition for a hundred metronomes called *Poème symphonique* (Symphonic Poem, 1962). In this piece he uses mechanical metronomes, whose ticking grows slower in varying tempos and which creates a complicated rhythmic micropolyphony.

Between 1985 and 2001, Ligeti wrote 18 etudes for piano[126] that are evidence of perfectly refined creative thinking. The etudes are modeled with different layers playing at once. These compositions, unlike Nancarrow's studies, can be performed live. For example, the etude *Cordes à vide*[127] (Open Strings) is typified by a polyphony of different metric accents, while the harmony is based only on the projections of the fifths. The etude *Automne à Varsovie*[128] (Warsaw Autumn) is exceptional for the fugue technique. An especially complicated rhythm motion is created, which creates the impression that the pianist is playing in 2, 3, even 4, separate tempos at the same time. According to Paul Griffiths, this piece's chromatic layers are analogical to Mandelbrot's computerized images (Griffiths/Ligeti

126 The cycle *Études pour piano* (Etudes for piano) is divided into three books: the first consists of etudes No. 1–6, the second – No. 7–14, and the third – No. 15–18.

127 Etude No. 2 is dedicated to the composer Pierre Boulez, from the first book of etudes, created in 1985.

128 Etude Nr. 6 is from the first book. Its title derives its meaning from the Polish modern music festival Warsaw Autumn, which has been organized since 1956. For a long time, it was the only festival of its kind in Central and Eastern Europe. The etude is dedicated to the composer's Polish friends (French *À mes amis Polonais*). Paul Griffith believes that with this etude the composer expressed solidarity with the Poles for Poland's difficult political and economic position (Griffiths/Ligeti 2001: 694).

2001: 694). The score to the left hand of the etude *Fanfares*[129] continually repeats the eight-tone series, which is made up of two ionian tetrachords from *c* and *f-sharp* and which is divided into three groups of eighth notes, 3 ♪, 2 ♪, and 3 ♪. The series is combined with the major and minor triads and seventh chords in the part for the right hand. The general sound creates the illusion of chaos in music because of a rhythmic conflict between the accents typical to the series 3–2–3 and the combination of eighth notes in the consonance of chords:

3–2–3–3–2   3–3–2–3–3   2–3–3–2–3   3–2–3–3–2

The title of the etude *Désordre*[130] (Disorder) encodes the acoustic chaos. However, the sense of disorder was created by the composer, holding on to constructively defined rhythmic sequences and a strict order. Typical of the composition is a rational blending of metro-rhythm and accents, as well as the drama of their exchange (tendencies of polymeter).

The analysis of the metro-rhythmic structure shows that the piece is compiled with accentuated and shifted structures that are based on sequences of certain rhythmic values. For example, in the first section of the etude (mm. 1–33) the left hand is continually playing the line of 8 eighth notes. Its perpetual motion is combined with the periodically shortening of this rhythmic group by one eighth note in the part for the right hand (shortening happens every fourth measure; see Figure 54). This results in displacement of metric accents (downbeats). After a certain time, at the end of the first section, the part for the right hand becomes a measure longer (in the right – m. 33 = in the left – m. 32).

The derived graphic scheme of the metro-rhythmic shifting of the first section I compared with the scheme of the third section. In this manner we come up with a polymeter palindrome that is formed from a distance, because in the third section an analogical shift of accents is made up in the part for right hand, where one eighth note is added every fourth measure (see Figures 55 & 56).

A certain constructive logic is also typical to the harmony of *Désordre*. The etude is composed only of two scales. The right hand part is modeled by nine tones of "white-notes" scale, while in the left hand part we hear nine tones of "black-notes" scale. Both right hand and left hand create a two-voice polyphony – the eighth note motion versus "hidden" melody in quarter and half notes. The schematic expression of both melodies shows that the tone scales are formed in a cone shape (see Figures 57 & 58).

---

129 Etude Nr. 4 is from the first book and dedicated to the German pianist Volker Banfield.

130 Etude Nr. 1 is from the first book and dedicated to Pierre Boulez.

Figure 54. Liget, Désordre. *Shift of metric accents in mm. 1–29*

Figure 55. Liget, Désordre. *Diagram of the first section, mm. 1–33*

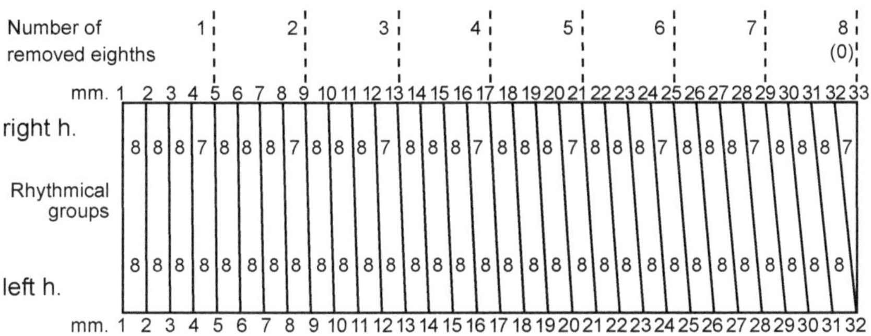

Figure 56. Liget, Désordre. *Palindromic diagram of the first and third sections*

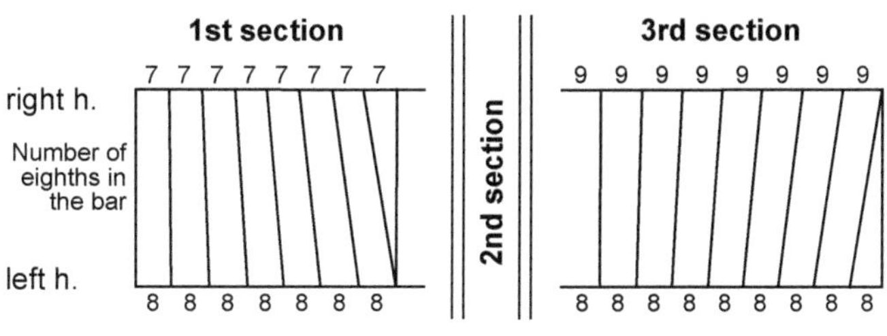

Figure 57. Liget, Désordre. *Right hand part: "white-notes" scale and diagram and organization of the melody*

Figure 58. Liget, Désordre. *Left hand part: "black-notes" scale and diagram and organization of the melody*

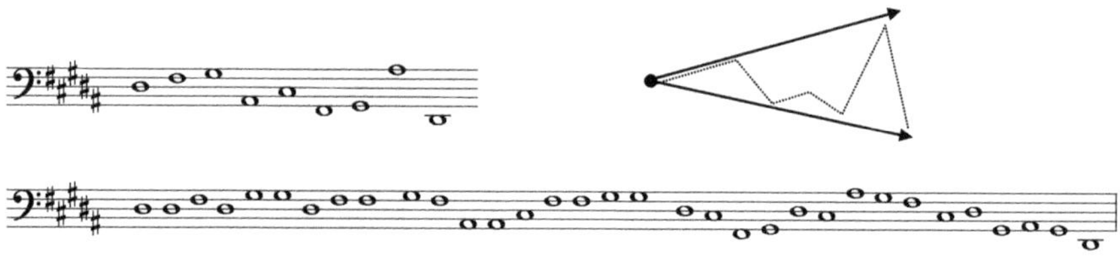

1.3. Symmetrical Algorithms and the Confrontation between Symmetries and Asymmetries

Hugo Riemann (1849–1919) discussed the expression of symmetry in the structure of music in his treatise *Katechismus der Kompositionslehre* (Catechism of Composition, 1889). He argued that "a strict symmetry and the opportunity for bilateralism is the essential foundation for all music forms" (Wille 1982: 12). The principle that symmetry is the foundational structure of all music was the basis Georgi Conus (1862–1933) used when popularizing his theory of metric analysis (metrotectonism, for example see Conus 1986). This is when the proportions of the groups of measures (a certain section of time) are viewed in the same manner one would view a symmetrical architectural construction. Additionally, the phenomenon of symmetry is researched as a universal coded origin, shaped by the musical materials of various epochs.

Larry Solomon, a researcher of the manifestation of symmetry in music, names the four permutations of musical material – O (original), I (inversion), R (retrograde) and RI (retrograde inversion) – as examples of rotational symmetry and supplements them by a Quadrate-variation method (diagonal transpositions; Solomon 1973: 260–1). The Quadrate-variation method gives composers more opportunities, because the combination of sounds, or series,

Figure 59. Traditional and diagonal transpositions (an example based and reproduced from Solomon 1973: 260)[131]

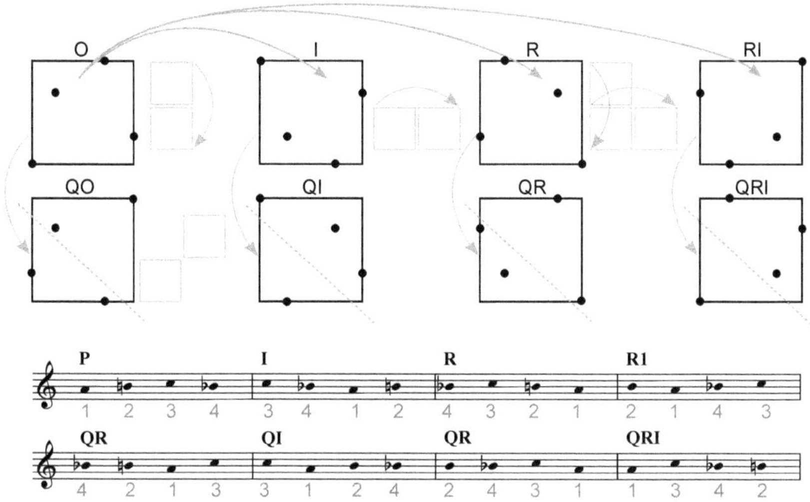

is not accurately repeated, but created out of modified forms distant from the prototype (see Figure 59).

The influence of symmetry in organizing musical tones, e.g., rotations of the same motif, is illustrated in the tonal relationships in Stockhausen's *Klavierstück III* (Piano Piece No. 3, 1952). The piece is composed from certain tone motifs, which are the transpositions and inversions of the six-tone model presented in the piece's first measures (Andreatta 2003: 129).

Symmetry in a musical composition can be expressed in the visual representation of a score or in the graphic prototype, opening up the logic behind choices of pitch. For example, Pozzi Escot (Escot 1999: 124) compiled the parameters for pitch in Ligeti's first etude for organ *Harmonies* (1967) into a figure in which the consistent process of cluster changing opens up like a symmetrical wave. Tom Johnson relied on his figures of mirror symmetry, which Jean-Paul Delahaye named "musical mathematical sculpture" (Delahaye 2004: 89), as the prototype for a 49-piece cycle for piano for four hands, *Symmetries* (1981–1990) (see Figure 60).

131 Rotation variants and tone indication (in grey color) supplemented by the author of this book – *R. P.*

Figure 60. Johnson, Symmetries. *Arrangement of the score according to the mirror symmetry (© Tom Johnson)*

More and more often in contemporary music algorithms of symmetry determine a composition's rhythmic image. For example, the Estonian composer Arvo Pärt (born 1935), in his piece for organ *Trivium* (1976), composes the rhythmic movement according to the principle of mirror reflection according to the number sequence 4–1–3–1–2–1–1–1–1–2–1–3–1–4. In Lithuanian music, we would set apart Balakauskas' so-called symmetrical sequence of relative tones in which the tones are laid out in fifths symmetrically from a chosen axis; mirror reflection is typical of the tone relationships. The symmetrical tone series are the source of sound in Balakauskas' composition *Auletika* (Auletics, 1966) as well as in the Second Symphony (1979):

d – a – g – e – c – b and *f –g-flat – b-flat – d-flat – e-flat – a-flat*

Even more, Balakauskas structurally forms the second parts of his cyclic opuses by reflecting the symmetry of an isomorphic series. We can see an example in his Second Symphony and in his *Dada concerto* for chorus and orchestra (1982).

Morton Feldman. ***Crippled Symmetry*** **(1983)**

In the oeuvre of the American composer Morton Feldman (1923–1987) there are a large variety of manifestations of constructivism. For example, in the cycle *Durations* (1960–1961) he manipulates the elements of the duration parameter. In the composition *Intervals* (1961) for an instrumental ensemble he employs the relationships of intervals. Feldman constructively nominated some of his compositions, e.g. *Vertical Thoughts* for various instruments (1963), *Numbers* (1964), *First Principles* for chamber ensemble (1967), and *Elemental Procedures* for soprano, chorus and orchestra (1976). In the third part of the monograph, I will explore more broadly the example of the original notation of the score of *IXION* (1958), which was colored with numbers.

The phenomenon of symmetry was also important in Feldman's creative space and was transposed in his composition *Crippled Symmetry* for flute, bass flute, celesta, glockenspiel and vibraphone (1983). By striving to create the effect of a broken symmetry in the tonal material of the composition, the composer uses symmetrical and asymmetrical models, which he refers to in his own words as "crippled."

Crippled Symmetry is noteworthy for its slowly changing musical flow. It is composed from prolonged or monotonously repeated tones and occasional rests that break into the composition. However, the composition's inner structure is based on rational calculations, which enable the broken symmetry idea. The analysis of the flute part displays an irregular repetition of musical proportions. A section of mm. 1–64 returns in measures 109–62 in the shape of an irregular retrograde. Only a 12-measure melody (mm. 109–21) is composed in a strict retrograde. Later, it does not retain the consistency of the retrograde. For example, the rhythmic figure 𝅘𝅥𝅯𝅘𝅥𝅯𝅘𝅥𝅯𝅘𝅥𝅯 𝄽 𝄽, returns in its original, and not mirror image, 𝄽 𝄽 𝅘𝅥𝅯𝅘𝅥𝅯𝅘𝅥𝅯𝅘𝅥𝅯. Furthermore, a few measures are eliminated (the retrograde section returns shortened – 54 measures instead of 64; see Figures 61 & 62).

The exact principles of symmetry are enabled in the central part of the composition. The graphic figure reveals that in mm. 361–406 the music takes on a typical mirror reflection (see Figure 63, upper section). However, later, from m. 406, the composer returns to the idea of "broken" symmetry, applies irregular retrograde, and brings in progressions of new duration (bottom section in Figure 63).

Figure 61. Feldman. Crippled Symmetry. *Diagram of a broken symmetry, flute part, mm. 1–64 and 109–62*

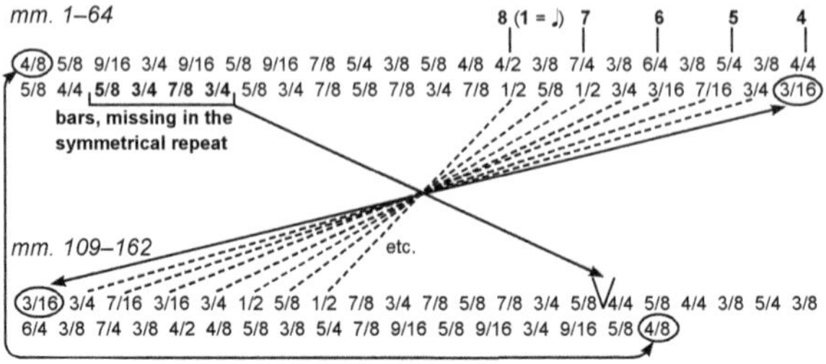

Figure 62. Feldman. Crippled Symmetry. *Score fragments of flute part, mm. 1–64 and 109–62*

Figure 63. Feldman. Crippled Symmetry. *Symmetrical arrangement of the central section (mm. 361–406), return of broken symmetry from m. 406*

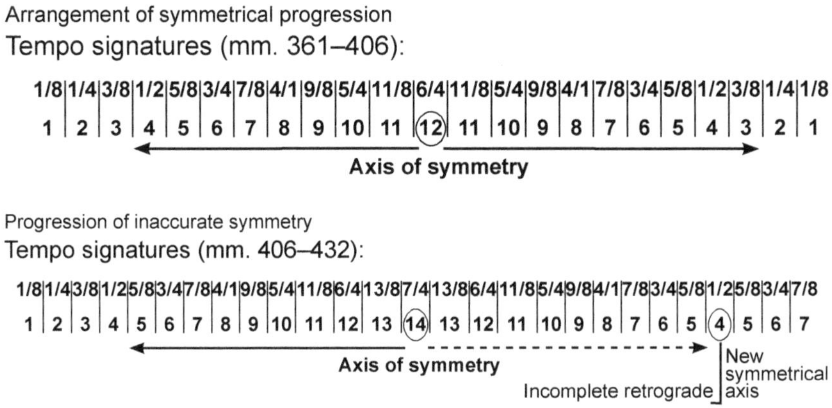

Jonathan Woolf argues that Feldman's inspiration for this composition were the paintings of Mark Rothko and the patterning of Oriental rugs, which at first glance appear geometrically accurate and yet are not in reality symmetrical (more see Woolf 2002). According to John Rockwell, Feldman tried to bring this principle to life by warping strict proportions, because "the sum of the parts does not equal the whole" (Rockwell 1999).

1.4. Transformational Elements (Combinatorics, Permutations, Rotations)

When analyzing combinatory actions applied to contemporary music, from which the parameters of a variety of tonal texture depend – pitch, rhythmic units, polyphonic segments, and so on, it is possible to generalize the direction of two combinatory processes as follows:

1) examples of traditional combinatory actions,
2) individual manipulations in composing, which have the features of combinatorics.

Combinatory actions were especially helpful in serial music. Combinatorics enable ways of transformation when using individual tones, allow the composer to rearrange rhythm, and allow new series to be developed. For example, Messiaen's technique of interversions (permutations) is a creation of variations from the same tonal series. One of the refined examples from Messiaen's breadth of creative work is the durational "tree" for etude *Mode de valeurs et d'intensité*, which was created from three series (Figure 64).

Figure 64. Messiaen, Etude Mode de valeurs et d'intensité. *Arrangements of three durational scales*

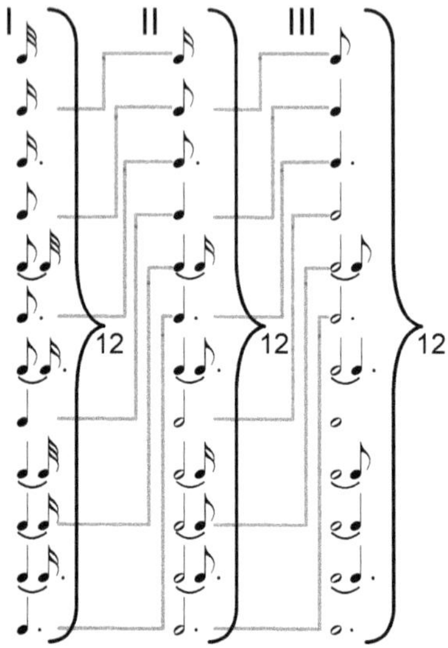

Olivier Messiaen. *Île de feu II* (1950)

Messiaen's widely use of pitch, articulation, and dynamics system, the so-called interversion method, becomes especially apparent in his etude *Île de feu II* (Fire Island II, 1950). The different principles of composing form two musical sections that switch back and forth. One of the two is based on interversion logic. In the score, the interversions are marked with the notations *Intervers. I*, *Intervers. II*, and so on. They are exceptional for their strict compositional logic and are organized according to the following process: from the original tone series the first variation, the interversion, is created through chance. The second interversion is created by applying that same action, which is no longer based on chance (mathematically this corresponds to multiplying a permuation, as an element of the symmetric group S_{12}, with itself). The technical interversion method is shown in Figure 65 with the number sequence from 1 to 12 along with two more sequences developed from it.

Figure 65. Above: the technical example of interversion method; middle and below: Messiaen, Etude Île de feu II. *Structural scheme of the etude and score fragment of interlude A*

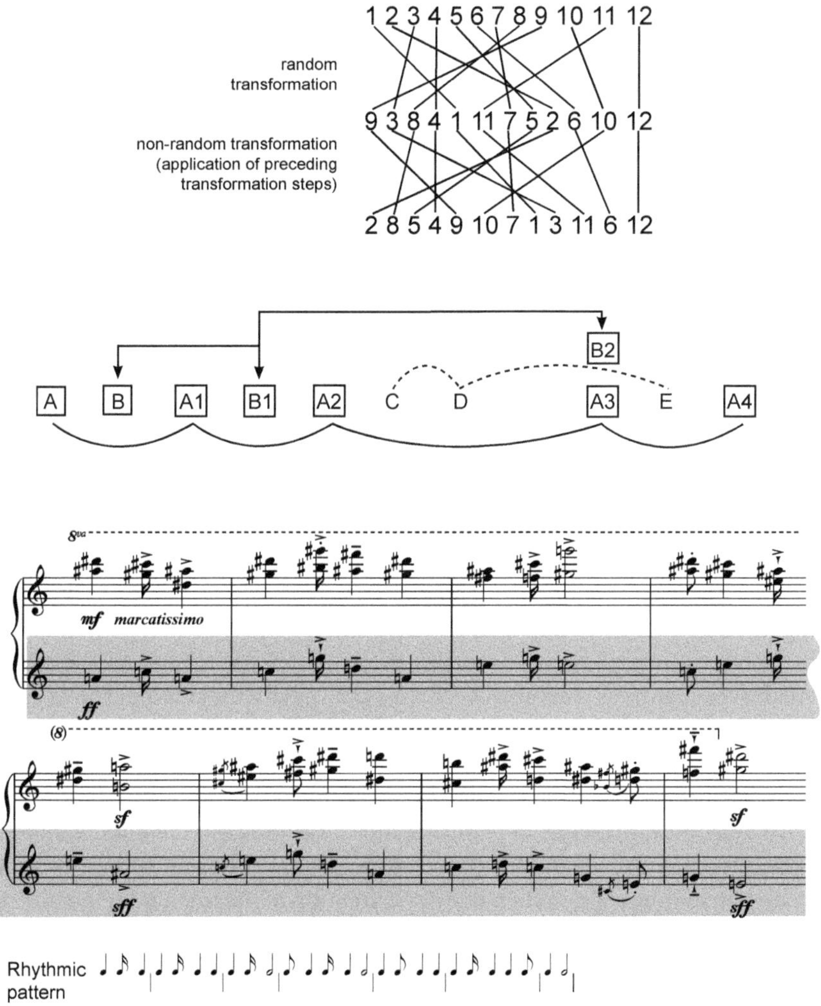

In the etude *Île de feu II*, in total Messiaen used ten intervesions (in Figure 65 for the interversion block, the letter B and its variations B1 and B2 are used). The interversions were combined with other musical sections – interludes, which are variant repetitions of the same material and which maintain the same rhythm (these interludes are marked with alphabetical equivalents A, A1, A2, A3, and A4). Besides, the etude was enriched with the sections C, D, and E that mark the new musical material.

In the etude's sub-heading, Messiaen presents an example of the original series, which is laid out according to the chromatic scale downwards from *b* to *c*. This is a model of total (integral) serialism – pitch, time, dynamics, and articulation are all organized together into one common system. Ten variations can be heard in the etude, making it an interversion prototype. The composer also applies total serialism in the interversion sections. When constructing interversions, simultaneously the dynamics, articulation, and duration that "belong" to each tone is permutated.

The first interversion is made from the center of the prototype. A symmetrical expansion begins with the two tones *f-sharp* and *f*, alternately selecting the tones up and down (see Figure 66). The second interversion is created by applying the same principle. This time the center is established by choosing the middle tones of the first interversion, *e-flat* and *a* (see Figure 67, section 1). The second interversion is presented in the left hand's part and is played together with the first interversion. By the way, an internal law was noticed in the structure of the second interversion. This is a tritone chain that is combined with the intervals progressively increasing from the minor second to the perfect fourth (see Figure 67, section 2). When I put together a graph of the progression from the center moving in separate lines, connected with a unified scale curve, I came up with a graph of three curves crossing over each other (see Figure 67, section 3).

An analogical principle of choosing tones was used to create the remaining eight interversions. Each time the central axis was chosen as the preceding scale's two middle tones. When one pays attention to the tenth interversion, one sees a scale-prototype retrograde, its central tones become the starting point for the inversion of the first interversion (see Figure 67, section 4).

Figure 66. Messiaen, Etude Île de feu II: *1) chromatic tone scale – prototype; 2) arrangement of interversion from the center tones* f-sharp *and* f; *3) tone scale of the first interversion; 4) visualization of the first interversion; 5) musical fragment of the first interversion*

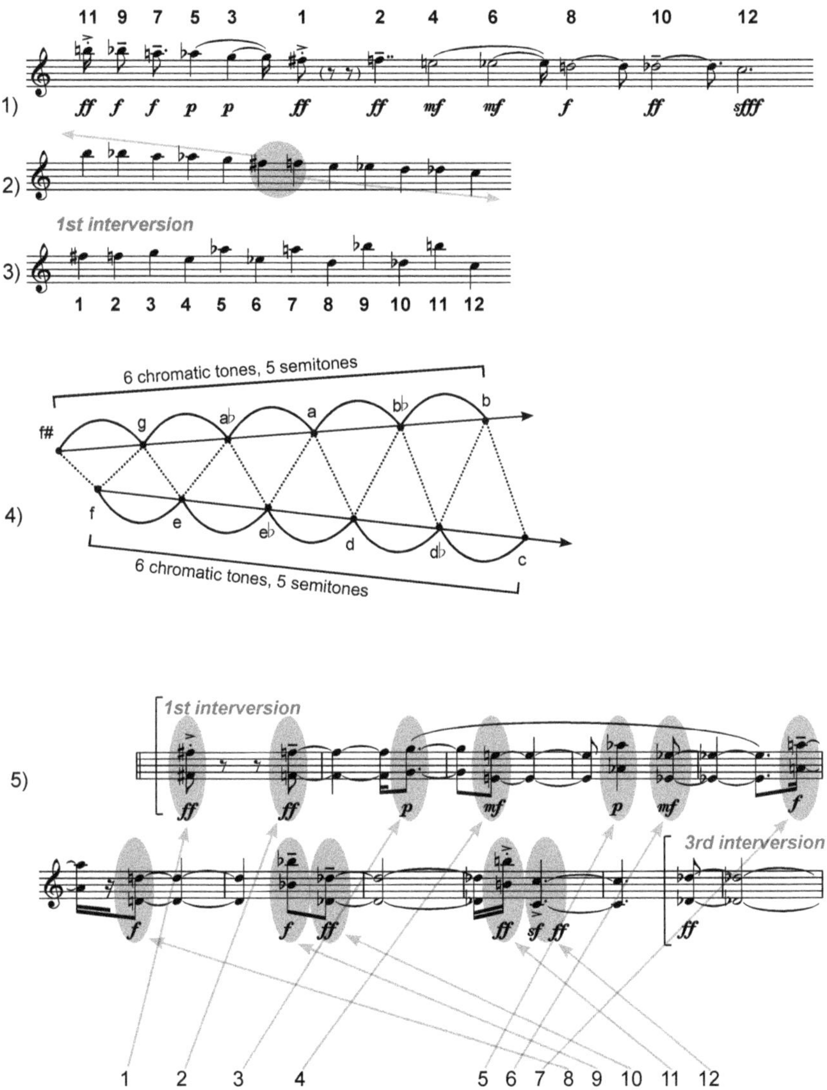

Figure 67. Messiaen, Etude Île de feu II: *1) arrangement of the second interversion from the center tones* e-flat *and* a; *2) interval features of the second interversion; 3) three intersecting curves in the structure of the second interversion; 4) arrangement of 3–10 interversions*

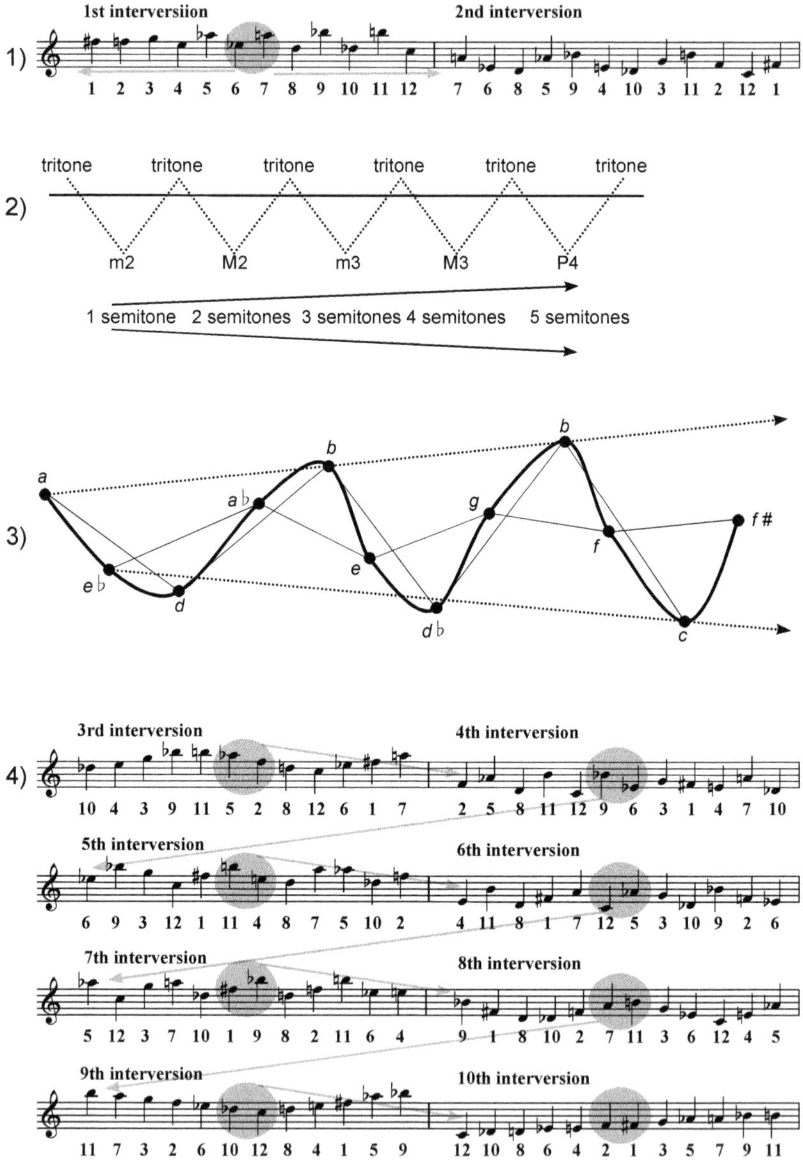

Tom Johnson. *Tango* (1984)

Combinatory actions are particularly common in Johnson's compositional practice. For example, in the cycle of pieces for various instruments *Tilework* (2002), rhythmic permutations are on display. Different rhythmic models are combined in such a way that every unit is filled, but two rhythmic units would not sound at the same time.[132] In the composition *Music and questions* for bells or glockenspiel (1988) the musical pattern is made up of 120 possible permutations of a five-tone motif.[133] The combinatory principle is used to compose the piece for player piano *Full Rotation of 60 Notes through 36 Positions* (1996). An analogical action is used in the vertical construction of the composition *Voicings* (1984): four pianists repeat the same phrase; however, in the vertical alignment of the score the music lines are constantly shifting.

Johnson's piece for piano *Tango* (1984) illustrates the use of traditional combinatory actions. This piece could be called a tutorial in musical combinatorics. In the composition all possible combinations of the five-tone motif d^2 – f^2 – *g-sharp*^{2} – a^2 – *b-flat*^{2} are used, in total 120 combinations. All of these combinations (each one is allocated one measure) are performed by the right hand, accompanied by the left hand's ostinato *D* – *A* – *d* – *A*, according to the Habanera fashion.

Johnson holds onto stability in his combinatory actions. He changes only the third, fourth, and fifth tone. He does not change the first two tones. For example, in the first line of the score, the tones d^2 and *f-sharp*^{2} do not change. In the second line, *d* and *g* do not change and so on (see Figure 68). Having used this method to create six versions (because 1 x 2 x 3 = 6), the combinatory action begins again from the beginning with another stable two-tone motif. This way, from the tone d^2 onwards 24 combinations are played. Having used all the variations onwards from the tone d^2, the composer brings forward the second tone (as the first tone of the motif) and applies an analogical action; then the third, and so on. Through the procedure of a lexicographical order, the music material consistently fills up with 120 five-tone motif versions (24 x 5 = 120).

132 This is an application of mathematical phenomenon – the Vuza canon, which is discussed in detail in the third part of this book, pp. 236–237.

133 Five elements can be combined into different variations, based on the variety applied through the principle: 1 x 2 x 3 x 4 x 5. In total, we get 120 permutations.

Figure 68. Johnson, Tango. *Combinatoric manipulations*

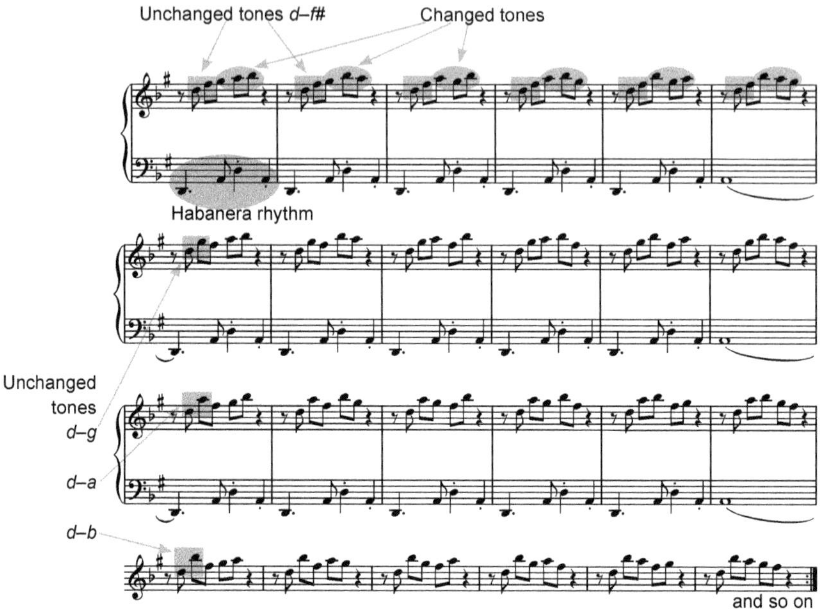

Tom Johnson. The Chord Catalogue (1986)

Another of Johnson's piece for piano, *The Chord Catalogue* (1986), solves a combinatory riddle. The riddle is how to arrange thirteen chromatic tones (c^1 – c^2), systematically organizing all possible sound combinations from two to thirteen tones, making up a sequence of all possible chords, and by so doing, elementarily filling the chromatic scale (see Figure 69). The composer calls each smallest two-tone derivative a "chord." Therefore, a more accurate description would be a piece made up of sound combinations (or even clusters). According to the rule of combinatorics, there is a total of 8,192 (2^{13}) possible combinations that can be made from the elements ranging from one to thirteen. However, the score is made up of 8,178 sound combinations, because the composer did not use an empty chord and 13 variants of one sound.

The harmony of *The Chord Catalogue* has been described as one persistent chromatic shift, until all the possible chosen types of chords have been used (two-tone, three-tone, four-tone combinations and so on). A graphic chart of the section where all possible three-tone combinations are played with g^1 held above exposes a self-similar principle that had been applied, because of the graphical "growth" of the chromatic shift (see Figure 70). Second, when I counted how many sound combinations there were from each type of combination (two-tone, three-tone combinations and so on), and after I put together a graph, it became clear that the laws of numeric repetition and the combinations of numbers were set up according to Pascal's triangle logic[134] (see Figure 71).

Figure 69. Johnson, The Chord Catalogue. *Organization of 2-tone and 3-tone chords*

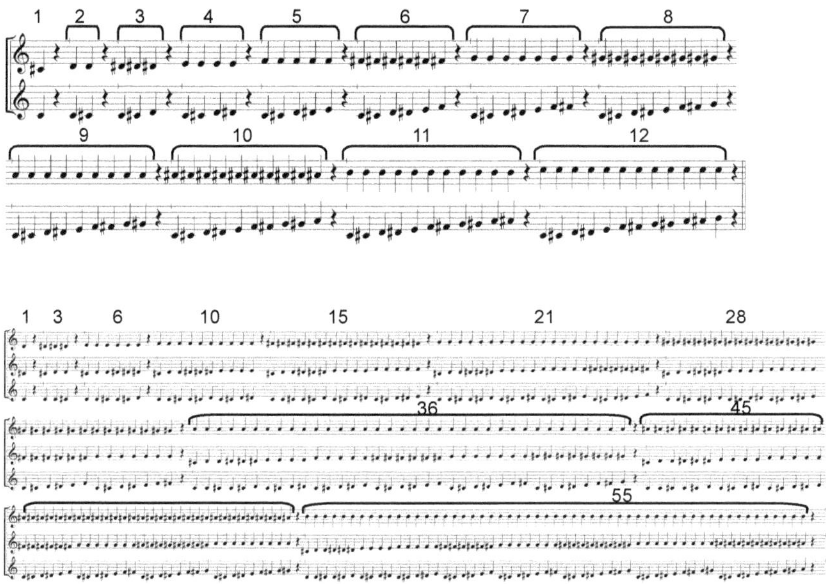

134 In the composition, the number of chords created from each tone reflects the number sequence from Pascal's triangle rule. This triangle is created by applying the formula $a_{nr} \equiv \frac{n!}{r!\,(n-r)!} \equiv \binom{n}{r}$, or every element is achieved by adding together the two elements above them.

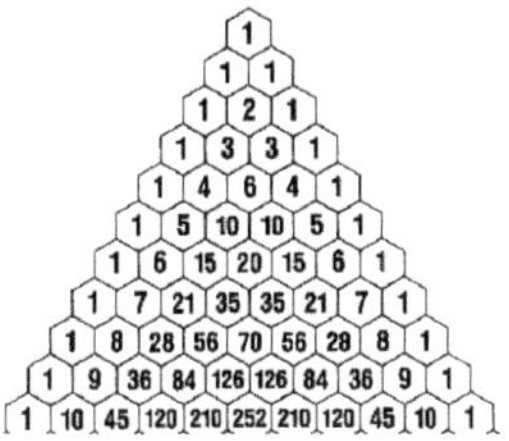

and so on

Figure 70. Johnson, The Chord Catalogue. *Diagram of 3-tone chords*

Figure 71. Johnson, The Chord Catalogue. *Numerical reduction of the musical score and the manifestation of Pascal's triangle*

Symmetrical number rows

2-tone chords	1	2	3	4	5	6	7	8	9	10	11	12	/ 78
3-tone chords		1	3	6	10	15	21	28	36	45	55	66	/ 286
4-tone chords			1	4	10	20	35	56	84	120	165	220	/ 715
5-tone chords				1	5	15	35	70	126	210	330	495	/ 1287
6-tone chords					1	6	21	56	126	252	462	792	/ 1716
7-tone chords						1	7	28	84	210	462	924	/ 1716
8-tone chords							1	8	36	120	330	792	/ 1287
9-tone chords								1	9	45	165	495	/ 715
10-tone chords									1	10	55	220	/ 286
11-tone chords										1	11	66	/ 78
12-tone chords											1	12	/ 13
13-tone chords												1	/ 1

Total 8178

In general, it could be said that arranging all the possible sound combinations in the chromatic scale, as was done in this composition, had the more likely outcome of creating a consistent mechanical action, rather than a work of art. However, the musical result excelled the elementary background. According to Kyle Gann, in the process of creating chords one can hear a retrospective of musical harmony from linear thinking and functional tonality to the epochs of seventh chords and ninth chords or analogies of serial music (Gann 1987: 84). Matthias Entress described the *The Chord Catalogue* as one more "dizzying multiplicity of colors present in the equal-tempered scale", "[a] transcendental experience [...] that well surpassed the simple combinations game" (see Entress 1998).

2. Semantic Aspects of Music Composition

The constructive nature of composing contemporary music often correlates with a semantic element, that is, with the rendering of notional underlying implications to a musical composition. Then a certain mathematical phenomenon or a formula, number or a complex of numbers in the musical material is used by a composer both as a working tool and as a symbol accumulating a certain meaning as an all-embracing idea or the contents of a composition. Therefore I raise not only the question "how" or "in what way" a specific number is coded and conveyed in the musical score, but also "why" the texture of music is given a meaning and "what" kind of message is the musical composition.

Different numerological practices of the earlier epochs are applied in the music of the 20th–21st centuries both in a traditional and individual way. In assessing the character of a semantic composing, several tendencies generalizing the authors' ideas can be distinguished. They are as follows:

- the application of cosmological numeral codes and graphic constructions, which render symbolic meanings of the cosmic structure to the musical architectonics;
- the implementation of kabbalistic symbols (e.g., magic square of numbers) and esoteric principles of the magic of numbers, cases of their use;
- the implications of sacral numbers, especially including the examples of the expression of Christian numerology;
- the selection of personalized numbers as constructive elements. This illustrates a trend for individuality that is especially characteristic of contemporary authors who like to structure musical writing through numbers that generate individual meaning.

2.1. Cosmological Number Codes and Graphic Constructions

The manifestation of cosmological number codes, which are based on an established number-symbol system, can be found in musical scores, such as equivalents of the seasons of astronomical years or an interpretation of the cosmic structure. The musical composition can be perceived through the antique Harmony of the Spheres; this is when the prototype is drawn from the symbols of a heavenly body (the stars, planets) and its numerical equivalent. The numerical signs of the Zodiac or the meanings of the astronomical year may be carried into the score of the music. In addition, the musical composition may be arranged graphically like a construction of symbols of cosmological bodies.

Among the 20th–21st century composers, Karlheinz Stockhausen often exploited cosmological themes in his compositions. According to Wager (Wager 1998: 158), the Sirius Star was especially significant in this composer's work.[135] This is connected with his biographical numerology. The composer knew that his birth date, August 22, was during the Dog Days; at this time Sirius becomes visible in Europe. This star inspired Stockhausen to compose his composition *Sirius* for electronics, voice, trumpet and bass clarinet (1977). The composer linked the tonal structures of this piece with the four seasons of the year, the four cardinal directions, the days of the week, the months, and the year. The composer specified in his notes that the four performers must sit in four corners, indicating the four directions. On the final page of the score, we come across an inscription of the number of the cycle of the Great Year (or Platonic Year). Symbolically, at the end of *Sirius*, the music acquires a quotation from the parts *Pisces* and *Aquarius* from Stockhausen's cycle *Tierkreis* (Zodiac, 1975). It may be surmised that with these quotes the composer symbolically defined the crossroads of two millenniums in sound – the beginning of the 2000s, when, according to some sources (including Wager 1998: 109), the astrological age of Pisces ended and the age of Aquarius began (this happens due to axial precession – a change in the orientation of Earth's rotational axis).

We may analyze Stockhausen's cycle *Tierkreis* for 12 musical boxes as the manifestation of the number 12 and an inlay of the zodiac. The composition is called "a cycle of musical formulae for the 12 months of the year and the 12 human types" (Moritz 2002). Among Stockhausen's other musical compositions based on cosmological themes are his allegory of the seven days of the week *Aus den sieben Tagen* (From the Seven Days, 1968) and his cycle of seven operas, *Licht: Die Sieben Tage der Woche* (Light. The Seven Days of the Week, 1977–2003).

Dmitri Smirnov presents a personalized interpretation of zodiac in his cycle *Two Fragments* for double bass, Op. 110 (1976–98). This piece is a phonic reflection of his and his wife's, Elena Firsova's, zodiac signs. The diptych is made up of two pieces. The first piece, *Scorpio*, bears the postscript "November 2", which is the composer's birthday (2 November 1948), and the zodiac sign Scorpio. The piece is made up of only one-tone and two-tone motifs. The second piece, *Between Pisces and Aries*, has the postscript "March 21." It is connected to the composer's

135 Sirius is the brightest star (Greek *Σείριος* – shining, hot), which is also part of the constellation Canis Major.

wife's birth date (21 March 1950) and the zodiac sign that is on the cusp between Pisces and Aries.

In Smirnov's composition *The Music of the Spheres* for piano, Op. 86 (1995), like in Paul Hindemith's (1895–1963) symphony *Die Harmonie der Welt* (The Harmony of the World, 1951), one sees the influence of cosmological semantics. This musical composition could be interpreted as the musical echo of the universe's architectonics. The number seven, which is the Antique symbolic number of planets, is also associated with Gustav Holst's (1874–1934) suite for orchestra *The Planets*, Op. 32 (1914–7). This is evident in the seven-part construction of the piece. However, it is necessary to note that in Antiquity the Sun and the Moon were among the seven heavenly bodies, while in Holst's suite there are encoded seven planets that were known during the composer's lifetime (including Uranus and Neptune, which were discovered in the 18th and 19th centuries), excluding the Sun and the Moon.

The idea of making the heavenly bodies into music is reflected in the work of George Crumb (born 1929). This is illustrated in the cycle *Celestial Mechanics*[136] for amplified piano, four hands (1979). The composer called this piece a suite of four cosmic dances. The title of the dance was chosen from the symbols of different constellations of stars. The composer arranged the order of dances according to the apparent magnitude of the stars, organized by the direction in which the light grows dimmer (1st piece *Alpha Centauri*, 2nd – *Beta Cygni*, 3rd – *Gamma Draconis*, and 4th – *Delta Orionis*).

The first two notebooks of Crumb's *Makrokosmos* (1972–3) are covered in cosmological semantics as well. They consist of 12 pieces bearing the names of the 12 zodiac signs. Furthermore, these cosmological symbols are personified, because in each piece Crumb ties each sign to an individual born under that sign. According to the composer, his first impulse was to create a cycle in which the titles – the initials for the names would remain a secret. This is analogical to the idea of the British composer Edward Elgar's variations *Enigma*, Op. 36 (1899): in the variations the composer encoded the initials of his friends and family. Like with Elgar's case, Crumb's riddle was solved. In his first cycle, the sign of Aries is encoded with the name of David Burge (No. 10 *Spring-Fire D.R.B.*). The composer encoded himself under the sign of Scorpio (No. 5 *The Phantom Gondolier G. H. C.*). In a similar fashion, he encoded the signs and initials of, among oth-

136 The cycle *Celestial Mechanics* in the composer's work is also called *Makrokosmos No. 4* from the series of cycles with the unifying theme *Makrokosmos*. The first two cycles, *Makrokosmos No. 1* and *Makrokosmos No. 2*, were written for piano. The third, *Music for a Summer Evening*, was written for two pianos and percussion.

ers, Federico Garcia Lorca, Brahms, and Chopin. Besides, the connection with Chopin can be heard in a few passages quoted from Chopin's *Fantaise-Impromtu* in C-sharp minor, Op. Posth. 66.

One more parameter is important to the subtext of these two notebooks: their graphic semantics and the visual presentation of their musical scores. These are expressed in specific geometric forms, which in the cosmological worldview take on a semantic shade and transfer their meaning into the space of music. For example, the fourth, eighth and twelfth pieces of the first and second parts of the *Makrokosmos* are given a graphic appearance through cosmological circles and spirals. An image of the passage of time in the form of a spiral makes a significant contribution to the finale piece of *Makrokosmos No. 1* – piece No. 12 *Spiral Galaxy [Symbol] Aquarius*. The score of the eighth piece under the sign of Leo, *The Magic Circle of Infinity [Symbol] Leo*, contains graphic symbols that depict the perfect Ancient Greek circle as a form of the cosmos (universal). The circle graphic is also typical of Crumb's vocal and instrumental composition *The Star Child* (1977, 2nd edition 1979).

Cosmological symbols as graphic prototype of a score can be found in the compositions of other composers. A good example would be compositions of John Cage: *Atlas Eclipticalis* for any 86 instruments (1962), which was composed on paper with an image of the astronomical map. The way in which the piano part was written in the etude cycle *Etudes Australes* for piano (1974–5) and *Etudes Boreales* for cello and/or piano (1978) was influenced by the positioning of the stars, because on his sheet of paper the composer noted the musical pitches in the places where there were stars on the astronomical map.

Graphic cosmological symbols are present in some contemporary Lithuanian music as well. These symbols appear in the work of the composer Bronius Kutavičius (born 1932), who is considered to be the pioneer of Lithuanian minimalism. They are visible in the pantheistic oratorios *Paskutinės pagonių apeigos* (The Last Pagan Rites, 1978), *Iš jotvingių akmens* (From the Yotvingian Stone, 1983), and *Magiškojo sanskrito ratas* (The Magic Circle of Sanskrit, 1990; see Figures 72 & 73). We could interpret the notation at the score of *The Magic Circle of Sanskrit* as directing the musicians to sit in a circle as they perform this piece as an allegory to the cosmic circle.

This Lithuanian composer has also used the symbols of astronomical numbers in his musical practice. The construction of his piece *Ten toli, iki vidurnakčio* (Far Away, Till Midnight, 1995) was inspired by 24 Indian modes, which the composer compared with each hour of the day. The modes symbolically repeat the time of day until sunset. Kutavičius' string quartet *Anno cum tettigonia* (Year

Figure 72. Symbols of sun and circle in Kutavičius' musical scores of The Magic Circle of Sanskrit

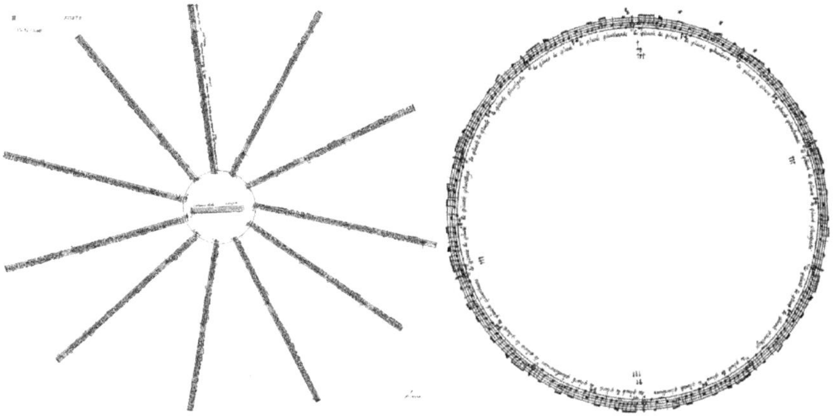

Figure 73. Symbol of the star in Kutavičius' musical score of The Last Pagan Rites

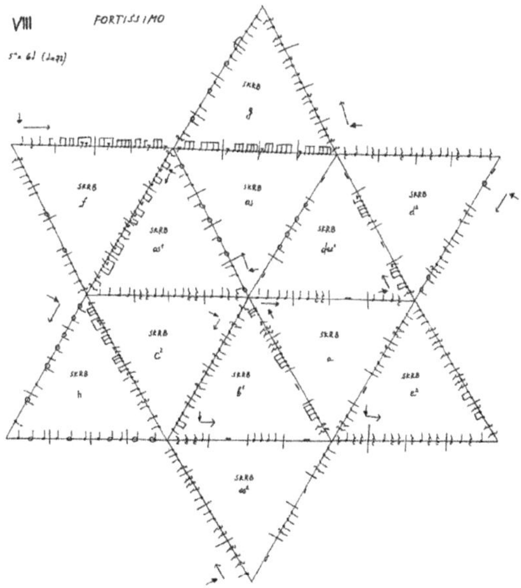

with the Grasshopper,[137] 1980) could be called an example of rational composing of music. Its prototype are numerical symbols of years, months, weeks and days, which influence the composition's macrostructure and structural microelements, e.g., division into sections, measure groups, number of tones. Therefore, the use of the word "grasshopper" in the title of the composition is not accidental (e.g., in Japanese mythology the grasshopper is the symbol of the year). For example, for the volume of the piece, the composer chose the unit of the day and accordingly composed the quartet in 365 measures (there are that many days in the year). Every 7 measures (a week) a new tone is introduced, and the rhythmic formula is developed upon. In total, in the composition 12 tones are played. The symbolic end of the month is emphasized with the ringing of a bell every 30 measures. The organ playing four tones symbolizes the four seasons of the year. The development of music and the tonal progression is continued into the climax of the score, which symbolically coincides with the date of solstice.

2.2. The Symbols of Magic Number Squares

The way that contemporary composers manipulate the magic number square when selecting durations, parameters of pitch or dynamics are not only constructive actions. They are supplemented by the composer's intent to compose so-called magical musical structures. One of the most influential examples of a magic square is the palindrome:

SATOR-AREPO-TENET-OPERA-ROTAS

The palindrome influenced many composers to employ the principles of the square to sort out the relationships of tone pitches and to determine their duration.[138] An anthologized example is Anton Webern's (1883–1945) Concerto, Op. 24 (1934) and his tonal series square. After Webern's sketches were published, it became clear that the composer modeled his work's series based on the Latin word-square.

The numerical analogy of this square is a transcript into the numbers 1 through 8, which correlate with the eight different letters of the literary palin-

137 *Anno cum tettigonia*, String Quartet No. 2, was composed by commission by Krzysztof Penderecki for the 1980 Warsaw Autumn Festival.

138 The possibilities of this Latin word-square are intriguing also because out of all the letters used, S, A, T, O, R, E, P, and N create the word combination "Pater noster" as it is laid out in a Greek cross (Latin *crux immissa quadrata*). The cross is made up of two lines of equal length, which cross through the center of both. Along their

drome. The composer John Tavener (1944–2013) applied these to his composition *The Protecting Veil* (1988). The number eight gave meaning to the composition's structure, which was made of eight parts. It gave meaning to the selection of eight different tonal centers. The so-called *veil theme* is made up of two sections of eight tones each. The *tableau theme* is made of a motif of eight tones; the manner of their permutations recreates the process of the palindrome (Tatom 2000: 7). The logics of the string quartet *The Hidden Treasure* and the oratorio *The Resurrection* (both composed in 1989) are based on an analogical method, which Tavener called "The Byzantine Palindrome." The logic of the magic square comes out in the tonal structure of his composition *Akathist of Thanksgiving* (1987), which was based on orthodox prayers and hymns of thanksgiving.

Judging by his initial constructive creative sketches, it could be surmised that Luigi Nono (1924–1990) was intrigued by the compiling of number squares. In his compositional practice he used square models as an expression of his total serialism, organizing the tonal pitch, duration and dynamics. To achieve this the composer used 12-number squares, which corresponded to the number of chromatic tones. When organizing the variety of dynamics the composer also used 12 different dynamic indicators and their combinations. Magic is not typical of Nono's squares. But the laws of changing elements are applied: a new line in a square begins from the second member of the former line. Using this principle, the dynamic relief of the first and second part of the concert for violin and chamber orchestra *Varianti* (Variants, 1957) was created. This principle also influenced the parameters of duration and dynamics in the oratorio *La terra e la compagna* (The Earth and the Companion, 1957). Besides, Nono used another way for number transposition: first–last, second–second to last, and so on. According to Erika Schaller (Schaller 1997: 131), this principle is applied to create the 12 series in his cantata *Il canto sospeso* (The Interrupted Song, 1955–6).

---

edges two letters remain, A and O, which symbolize the beginning and the end, "alfa" and "omega":

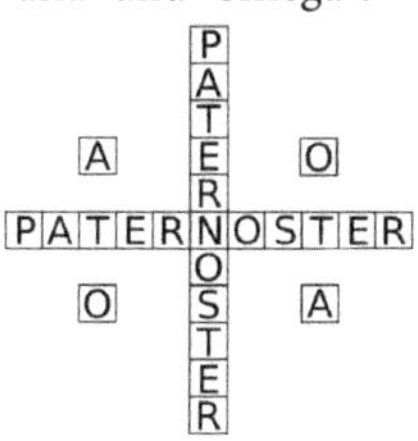

Šarūnas Nakas. *Ziqquratu* (1998)

Radical, often extreme contrasts (especially in terms of dynamics, timbre) are used in the compositions of Lithuanian artist Šarūnas Nakas (born 1962). Rationality is typical of his compositions. Mostly, the composer places his essential focus on their rhythmic arrangement. An example is his opus *Chronon* (1997): a multi-isorhythmic palindrome is formed in the first part of the cycle *Ištakos. Paukščiai* (The Origins. Birds); in the structure of the third part, *Marios. Dangus* (Lagoon. Heaven), the contour of an isorhythmic motet emerges; the movement of the second part of the composition, *Upė. Rėvos* (River. Shoals), is described as "a marathon that necessitates the precision of a jeweler" (Nakas 2002: 2).

In the structural plan of his composition *Ziqquratu* for flute, clarinet, violin, cello, piano, and percussion (Ziggurat, 1998), the rhythmic precision is expressed in various numerical combinations and in certain numerical symbols. This piece was commissioned for a festival in Switzerland. At first, it could not be performed because the rhythm was too complicated; the premiere took place a year later using electronics and was called *Ziqquratu-II*. An analysis of the score of *Ziqquratu* revealed that Nakas formed the complicated rhythmic counterpoints by strictly adhering to mathematical logic, based on selected numerical symbols, i.e., the structural idiosyncrasies of ziggurat pyramids[139]

Figure 74. Arrangement of squares of numbers 7 & 5

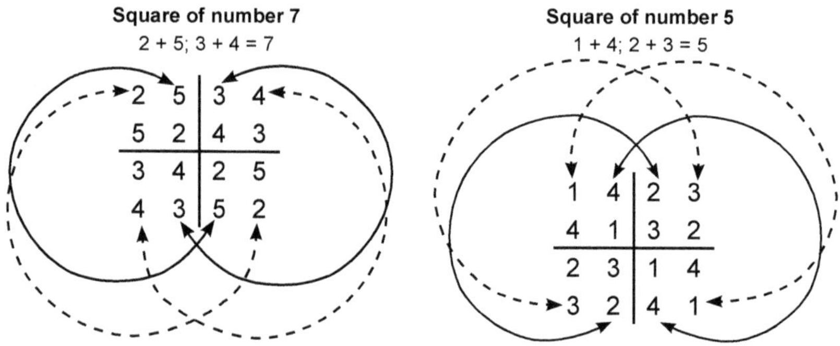

139 Ziggurats is a three- to five-thousand-year-old temple constructed in the form of a pyramid with stairs or in the form of a tower, which symbolized the Holy Mountain. Ziggurats are found in Sumerian lands, Assyria, and Babylon.

and the rules of the magic number squares. It was established that there were applied two squares with the roots of 7 and 5. The reason for manipulating them was that the composer most likely chose the length in letters of his name and surname (ŠARŪNAS – 7 letters, NAKAS – 5; see Figure 74).

Nakas used the number seven as a constructive compositional concept and as the ziggurat symbol. According to the composer, the early sketches of this composition had graphic representations of the ziggurat shape. The ziggurat is made up of seven tiers leading to the top. These levels correspond to the respective colors that encompass the seven metals and celestial bodies (celestial spheres).

The number seven is chosen as a respective symbolic subtext. This number, and its variations, constructively influences the various parameters of *Ziqquratu*. For example, seven instrumental parts form the music score. There are 14 different percussion instruments (i.e. 7 x 2). The tonal pitch is dependent on the overtone spectrum of tones *b-flat* and c. The distance between these pitches is a major ninth, or 14 semitones. The composer created tonal series out of these spectrums, selecting 28 harmonics (7 x 4). The series are divided into four sections of seven tones each. In total, seven different rhythmic units are used in the scores for violin and wind; one unit is longer than the other by a thirty-second note (see Figure 75). In the rhythmic structure of the piano score, four rhythmic series of seven members each are manipulated. An example of how number squares influence the rhythmic structure of *Ziqquratu* would be the melodic section of the cello part. So, I chose the value of the sixteenth note as a unit of measure and looked at how many of these units make up each one of the rhythmic motifs. Therefore, I determined that it was arranged according to the square of number seven[140] (see Figure 75, fragment of cello part).[141]

---

140 A detailed analysis of this composition was completed by me and was part of my Master's thesis, *Constructive and Symbolic Manifestation of Numbers in the 20th century Music* (Povilionienė 2003). Also, my analysis was published in the article "Music Composition as a Numerical Construction. The Projections of Magic Number Squares and Dimensions of Ziggurat Pyramids in the Composition *Ziqquratu* (1998) by Šarūnas Nakas" (Povilionienė 2006).

141 Further analysis of the composition *Ziqquratu* is presented in the section "Fractal Theory Analogies in Musical Compositions" of the third part in this book, pp. 243–245.

Figure 75. Nakas, Ziqquratu. *Seven rhythmic values, used in the parts of strings and woodwinds, and four rhythmic series; rhythmic design of cello part according to the number 7 square*

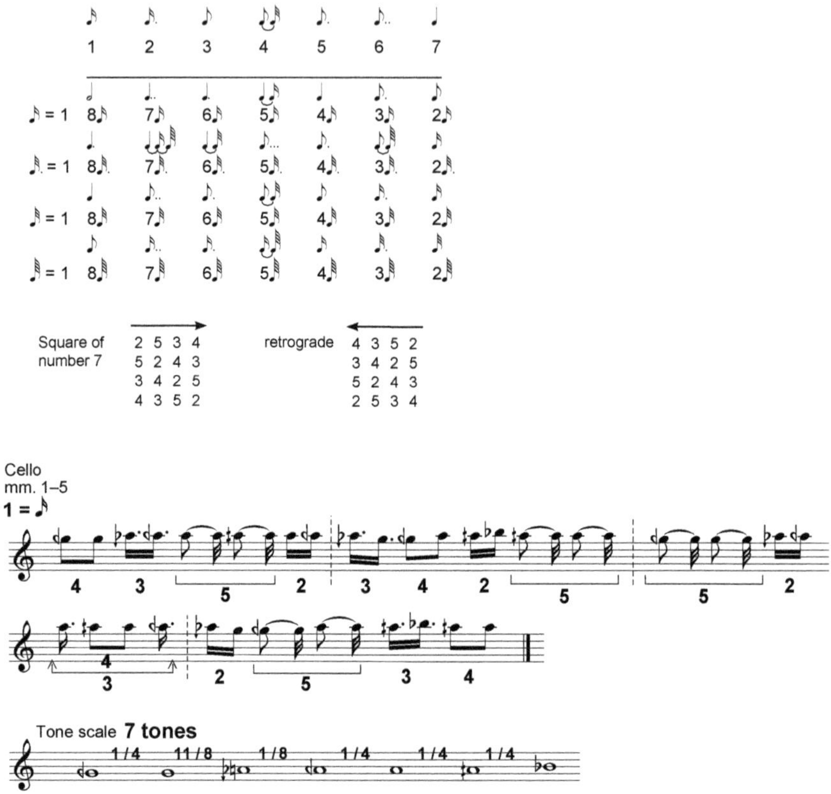

Dmitri Smirnov. *Two Magic Squares* (1971)

Smirnov's cycle of pieces for piano *Two Magic Squares* (1971) is an example of a composition, in which the rhythmic picture, and the organization of pitch, are strongly influenced by rules of the magical number square. For example, the first piece, *Magician,* is composed as all possible variants of the third row number square (vertically, horizontally, and diagonally). The scale, in mm. 1–3, is made up of 12 tones:

c – f – b – f-sharp – c-sharp – g – d – g-sharp(a-flat) – d-sharp(e-flat) – a-sharp(b-flat) – e – a

Figure 76. Smirnov, Magician. *Diagram of arrangement of square numbers; the number of tones in each measure, mm. 1–16*

In each measure, the square numbers determine the number of tones. After analyzing how many tones one can hear in each of the measures, I assembled the number sequence and reconstructed the process of how the square's sequence is applied (see Figure 76).

According to Smirnov, the inspiration for his second piece, *Bells*, was Albrecht Dürer (1471–1528) with his allegorical engraving *Melencolia I* (1514). In the right top corner of the work, the Renaissance artist etched a number square, which was often referred to as "especially magical" or "pan-magical" or even "devilish." This is because the sum of numbers in the square's horizontals, verticals, diagonals, small squares within the square, and even the broken diagonals always remain the same – 34. Additionally, it was noticed that the number pairs in the center of the last line match with the date the engraving was made, 1514 (see Figure 77). In Smirnov's piece *Bells*, the number sequences of Dürer's square influence the pitch and musical intervals. In the beginning the composer uses the large square's horizontal, vertical, and diagonal numbers. Later, from m. 22, he makes use of the numbers from the inner squares. From m. 35 there is an inversion of the inner squares' numbers. From m. 49 the inversions of the horizontals, verticals, and diagonals of the big square are used. Then, from m. 67, the numbers of the broken diagonals are used.

Dürer's engraving also inspired the German composer Michael Denhoff (born 1955) to write his piece for a small orchestra *Melancolia* (1980). According to the composer, a chain of triads in the middle section, that is noteworthy for its low register and dark orchestral timbre colors, emerged from a prototype of the number square. Denhoff created a musical square by setting the chords into the square's cells, whose vertical, horizontal, and diagonal were made up of a chain of four chords, which covered all 12 chromatic tones (see Figure 77).

Figure 77. Left: Dürer's Melencolia I; *right: Denhoff's musical square for* Melancolia

16	3	2	13
5	10	11	8
9	6	7	12
4	15	14	1

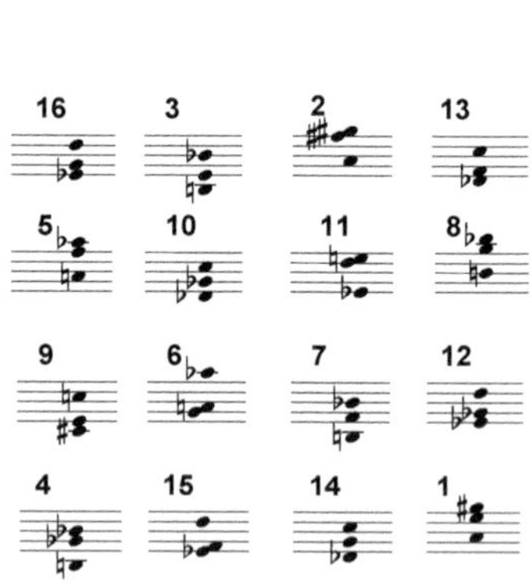

2.3. The Implications of Sacred Numbers

The numerological codes of Christian symbols in the compositional practice of contemporary music first and foremost are associated with an especially spiritual concentration in Messiaen's music, which stands out from the panorama of the 20th century. The truths of the Catholic faith were an inseparable part of the belief system of the French composer since childhood. He referred to the creative process as an "act of faith" (Bruhn 1997: 7) and brought this to life with certain symbols, which in his opuses can be observed as the marks of Christian numerology. For example, the number symbolism from the Book of Revelations is encoded in the composition *Couleurs de la Cité céleste* (Colors of the Celestial City, 1963). The sphere of the suffering of Christ is identified with the mystification of the sacred symbol of seven, which is found in *La Transfiguration de Notre Seigneur Jésus-Christ* (The Transfiguration of Our Lord Jesus Christ, 1965–9). This composition is divided into two large parts *septenarius*. The numerological semantics is applied to the number of orchestral instrument solos: seven solo instruments "chirp" bird figurations in the meditational episodes, which can be heard after the chorus gives comment of scenes from the Gospels (Tatlow & Griffiths 2016; also see Griffiths's article Griffiths/Messiaen 2001).

An especially moving way of creating sacred meaning within the texture of sounds is illustrated by Messiaen's cycle for piano *Vingt Regards sur l'Enfant-Jésus* (Twenty Contemplations of the Infant Jesus, 1944). The pieces become virtual symbols of the life of Jesus. Siglind Bruhn divides this cycle into four meaningful blocks: 1) the pieces No. 1–5 display main Christian symbols; 2) the sphere of the divine is represented by pieces No. 7, 9, 11, 13, 15, 17, and 19; 3) pieces No. 8, 10, 12, 14, 16, and 18 are interpreted as allegories for friendship, worship, and prophesy; 4) in addition, pieces No. 6 and 20 symbolize the synthesis of the creation of the world and the Christian Church as the "alpha" and the "omega." This meaningful division resulted in the symmetrical aspect of this cycle for piano. It is probably no accident that the most important episode of the divine theme is accentuated in the center. That is the piece No. 13 *Noël* (Christmas). From this axis in a symmetrical fashion two meaningful parts paired from the position of their similar terminology: No. 9 *Regard du temps* (Gaze of Time) and No. 17 *Regard du silence* (Gaze of Silence), No. 11 *Première communion de la Vierge* (First Communion of the Virgin) and No. 15 *Le baiser de l'Enfant-Jésus* (The Kiss of the Infant Jesus; see Figure 78) (Bruhn 1997: 87–8).

Figure 78. Messiaen, Vingt Regards sur l'Enfant-Jésus. *Symmetrical arrangement of compositions of the divine theme (No. 7, 9, 11, 13, 15, 17, and 19; based on Bruhn 1997: 87–8)*

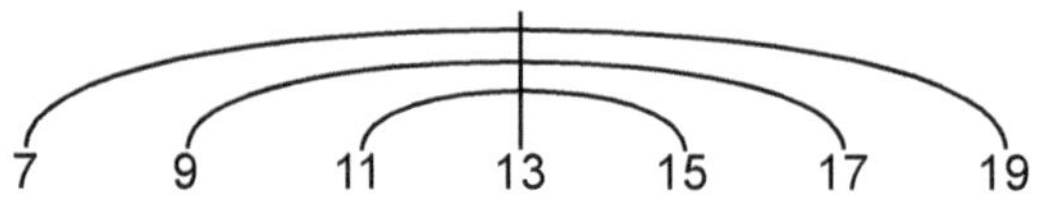

In *Vingt Regards* the composer probably consciously chose the numeration for the pieces based on Christian numerology: the number 1 from a Christian interpretation of the Ten Commandments (the Lord God) makes the first piece in the cycle meaningful, *Regard du Père* (Gaze of the Father). The numerical symbol of the Son of God is 2, which is an allegory for the stars of Bethlehem that led the way when Jesus was born. They are given special meaning in the second piece *Regard de l'étoile* (Gaze of the Star). The numeric symbol 5 represents Five Holy Wounds and can be seen in the fifth piece, *Regard du Fils sur le Fils* (Gaze of the Son upon the Son). The composer chose the sixth piece in the cycle for the tonal expression of the six days that it took to create the world, which is called *Par Lui tout a été fait* (By Him Everything was Created). The numeric symbol of the cross, 7, resonates with the title of the seventh piece in the cycle, *Regard de la Croix* (Gaze of the Cross). The numerological interpretation of the 12th piece, *La parole toute puissante* (The All-Powerful Word), could be explained as the numerical equivalent of the True Church.

In this cycle, there is also an analogy to the graphic cross, which is made meaningful through the relationship between musical tones. This can be detected in the so-called Theme of the Star and of the Cross (French *Thème de l'étoile et de la croix*),[142] which appears in the second piece (Gaze of the Star). The first four tones, *a – a-flat – b-flat – g*, visually open up into the sign of the cross (see Figure 79).[143]

142 This cycle is like an entire system of leitmotifs, which Messiaen calls the *Thèmes*. The composer wrote out the titles of the themes himself: *Thème de Dieu* (Theme of God), *Thème de l'étoile et de la croix* (Theme of the Star and of the Cross), *Thème de l'amour mystique* (Theme of Mystical Love), *Thème d'accords* (Theme of Chords), *Thème de danse orientale et plain chantesque* (Theme of Oriental Dance and Plainsong), *Thème de joie* (Theme of Joy) and *1er Thème* (1st Theme).

143 Already Bach recognized the contours of the cross in the tonal code of his own surname. The motif *b-flat – a – c – b* can be found in especially sacred parts of his compositions.

Figure 79. Messiaen, Regard de l'étoile. *Score fragment and graphics of the theme of the star and of the cross*

According to Wager, Stockhausen was a practicing Catholic. The numbers 7 and 12 in his music are associated with Christian meanings "as well as his daily life." For example, it is argued that Stockhausen considered it a significant coincidence that his 49th birthday was in 1977, because the number 49 is the square of the number 7 (Wager 1998: 88). According to the composer, the number 13 determined the musical changes in his cycle of operas *Licht: Die Sieben Tage der Woche.* When the opera's characters count to the number 13, just before this number the music takes on "all sorts of strange things" (Wager 1998: 87).[144] Moreover, Christian semantics are marked in the verbal texts used by the composer. In the composition *Gesang der Jünglinge* (Song of the Youths, 1956) 17 lines from the Book of Daniel are quoted; a mime participates in the composition *Inori* (1974), performing the gestures of delivering a mass in synchrony with the music.

One of the most meaningful numerical symbols in musical practice is the number three. This is the number of the Holy Trinity. This symbolic number is the basis, for example, of Smirnov's piece *Trinity Music*, Op. 57 (1990): it is scored for three instruments – the clarinet, the violin, and the piano. Tom Johnson's cycle of three anthems *Trinity* (1978) lends meaning to the number three not only in its title, but in its three-part structure, on a verbal level (the names of the figures of the Holy Trinity are chanted in English: *Creator, Son* and *Holy Ghost*), and in its minimalist sound structure: the piece is composed as the constant repetition of the three-tone motif *b – a – g*; its range creates the interval of major third.

The inlay of the symbol of the Holy Trinity can be seen in Onutė Narbutaitė's (born 1956) *Tres Dei Matris Symphoniae* for two choirs and a symphony orchestra (2003). The composition is made up of three parts: *Symphonia prima* "Angelus Domini" (according to Luke 1: 26–38), *Symphonia secunda* "Bethleem"

144 The musicians count evenly in Italian up to the number 13 (Italian *tredeci*) in part *Examen* (Exam) of the opera *Donnerstag, Luzifers Traum* (Thursday. Lucifer's Dream) or in section *Luzifers Abschied* (Lucifer's Farewell) of the opera *Samstag* (Saturday).

(according to Luke 2: 1–14) and *Symphonia tertia* "Mater Dolorosa" (according to Luke 23: 23–38, 44–46). They convey the three main moments of Mary's life – the Revelation, the Birth, and Death on the Cross. In the score and musical material of the work we find signs of the Holy Trinity as well. For example, in the first page of the score there is a drawing of the triangle. In the section *Stabat Mater* the seven *tutti* accents are repeated three times and are linked with the seven-note trombone motif, as though it were an allusion to the Seven Sorrows, and so on.

Bronius Kutavičius. *The Gates of Jerusalem* (*1991–5*)

Numerical manipulations of a sacred nature are typical of Kutavičius' cycle of oratorios *Jeruzalės vartai* (The Gates of Jerusalem, 1991–5). Already in the composition's title the word "gates" carries the encoded meaning of a symbolic transition from this world to the other world. The cycle's architectonics is determined by the number 12: it is made up of four separate oratorios (allegory of four cardinal directions), *Žiemių vartai* (Northern Gates), *Saulėtekio vartai* (Sunrise [Eastern] Gates), *Pietų vartai* (Southern Gates), and *Saulėlydžio vartai* (Sunset [Western] Gates); each of these is divided into three parts.

Respective numbers in this composition lend structure to the pitch and to the rhythmic relief. For example, the structure of *Southern Gates* is dominated by the number six. In the second part, the *ostinato* rhythm of the percussions is composed of rhythm values from one to six. The bassoon *ostinato* is played six times. Here the tones *g* and *a* are each repeated six times.

The structure of *Sunset Gates* is based on the number 5. The pedal point sounding in the second part of this oratorio is based on the five-tone motif *b – c-sharp – d – e – f*. In the third part the organ and double bass *ostinato* is extended through 15 quarter notes. *Sunset Gates* is given a sacred quality by the quotation of the Stabat Mater liturgical text. Therefore, symbolically *Sunset Gates* connects with another oratorio from the cycle, *Eastern Gates*, as they bring together the European Western and Eastern cultures. In the latter oratorio, the Japanese poet Yosa Buson's haiku are used along with the instrumentation of the Japanese gagaku.

The idea of the *Northern Gates* is seen as a turning back towards the pre-Christian, or pagan, rituals. Here, an archaic fish enchantment is repeated. Inga Jankauskienė argues that *Northern Gates* extends Kutavičius' favorite list of "ritual compositions" (Jankauskienė 2001: 71). A repetitive technique is typical to these compositions, as well as a steady pulsating rhythm, ostinato features, and the gradual lengthening or shortening of rhythmic values. While analyzing *Northern Gates* it was established that its structure is influenced by the number 9 (and its variant, 3). For example, the number 9 in the first part of *Northern Gates* influences a logics of rhythmic progression – the composer gradually extends the rhythmic values from one eighth note to nine quarter notes. Therefore, the rests grow shorter from nine eighth notes to one quarter note. In the second part 18 string instruments play at once (six first violins, four second violins, four violas and four cellos). They are divided into three groups according to how they enter. The tone scale is made up of the three-tone motif *b – d – e-flat*. The intervals of tone combinations *b – d* and *b – e-flat* are accordingly minor and major thirds. In this part the Karelian folk text that is quoted is a nine-syllable fish enchantment:

An -	*na -*	*pa*	*Ah -*	*ti*	*suu -*	*ri*	*hau -*	*ki*
1	2	3	4	5	6	7	8	9

(Finnish – Give me, Ahti, a big pike)

Rhythmically each syllable is matched to one eighth note. In the parts of each instrumental group, the enchantment is spoken in syllables according to a certain number sequence (this can be seen in the scheme and in the example of notes, Figure 80). The general compositional block of the enchantment is presented nine times in total. It has been established that the frequency of rests, which separate the rhythmic figures, depends on the progressive row: in the first five times repeating the entire compositional block the rhythmic figures are separated by one eighth-note rest. From the sixth repetition onwards, the duration of rests increases arithmetically: on the sixth repetition the rhythmic figures are divided by the rests of two eighth-notes; on the seventh repetition – the rests of three eighth-notes; on the eighth repetition – the rests of four eighth-notes; and on the ninth repetition – the rests of five eighth-notes.

Figure 80. Kutavičius, Northern Gates. *Rhythmic arrangement of the enchantment in the second part*

1st violins	9	1	7	6	5	3	4	2	8
1st violins	6	3	4	5	2	1	9	8	7
2nd violins	4	5	3	2	1	6	8	7	9
Violas	3	7	2	9	8	4	6	5	1
Cellos	8	2	1	7	6	9	3	4	5

Snieguolė Dikčiūtė. ***The Mystery of the Seven Bridges*** **(1991)**

The semantics of Kutavičius' *The Gates of Jerusalem* is made meaningful through the oratorio's view into Western and Eastern cultures, into the depths of the pagan and Christian worlds. All this is resembling the idea of Snieguolė Dikčiūtė's composition *Septynių tiltų misterija* (The Mystery of Seven Bridges, 1991). The symbolism of the bridge used in the title is interpreted as the road from one world into another, as the combination of what is inner and what is outer. The composer herself (Dikčiūtė 1993: 8) assigns this composition to sacral works (beside *Artava*, 1988, *Hosanna*, 1988, or *Mini Dogma*, 1990), because religious symbols can be found within the compositions. And the number 7 perfectly aligns the pagan and Christian cultures.[145]

An analysis of the composition allows me to argue that the semantics of the composition's title applied a constructive code that determines the organization of the musical material. The cycle is made up of 14 parts, which are divided into two sub-cycles of seven parts. The word "bridge" is used in the titles of the first sub-cycle parts. In the titles of the six parts of the second sub-cycle, the word "this side" is used, while the final part is titled "that side." The composition is scored for two instrumental groups with seven instrumental parts in each.[146] The choir is divided into two groups with seven voice parts in each.

The metric and rhythmic parameters also depend on the number seven. All the parts of the cycle are written in the 7/4 time signature. The beginning and the end bars in all parts are filled with seven quarter-note rests. The rhythmic texture consists of four rhythmic series, and each series is made up of seven rhythmic values.[147] The number seven is significant to the formation of the "white-notes"

---

145 In Lithuanian folklore the number seven is conceived as the symbolic number of the seven bridges. In Cosmology the number seven takes on meaning as the seven celestial bodies that create harmony in the heavens. In Christian numerology the number seven is the gift of the Holy Spirit. Seven is the number of deadly sins or the number of miracles. Other Christian meanings for the number seven are: faith, the church; the Apocalypse; the number seven represents virginity, innocence, seven virtues; there is peace on the seventh day after six days of creation; Jacob served for seven years, and so on.

146 The composer wanted to make the violin group consist of seven violinists. Therefore, she moved the eighth violinist to a group with other instruments (violas, cellos, and bass). In this manner she created the second group of seven: 1 violin + 3 violas + 2 cellos + 1 bass.

147 This aspect was discussed on pages 107–108 in this book, while presenting the application of a number progression to musical rhythm.

scale as well. In total, there are seven transpositions of the scale. Six transpositions from the tones *e*, *f*, *g*, *a*, *b*, and *c* are incomplete, each has six tones. However, in the parts IV and XII, the scale from the tone *d* is made up of all seven tones, *d – e – f – g – a – b – c*.

The composition's material was composed from the traditional series rotations and from combinations that at first glance did not seem to belong to the original series. In the work the "white-notes" scale is presented with five of its variations, generated by a technique of general rotation (a new series is made up of the tones that came before, which are laid out according to this order: first–last, second–second to last, third–third to last, and so on). However, in some places of the score we see the fragments of motifs that are different to the original series. Actually, these combinations are the result of an interesting manipulation – the secondary step of series rotation. The series and its primary (general) five rotations were written in one line. Then the composer chose rests on the boundary markers. In this manner, six new versions of motifs were derived (see Figure 81).

Figure 81. *Dikčiūtė,* The Mystery of Seven Bridges. *Original tone scale from the tone* e *and its rotations (according to this order: first–last, second–second to last, etc.) in the first part; diagram of secondary rotation process and six melodic variants*

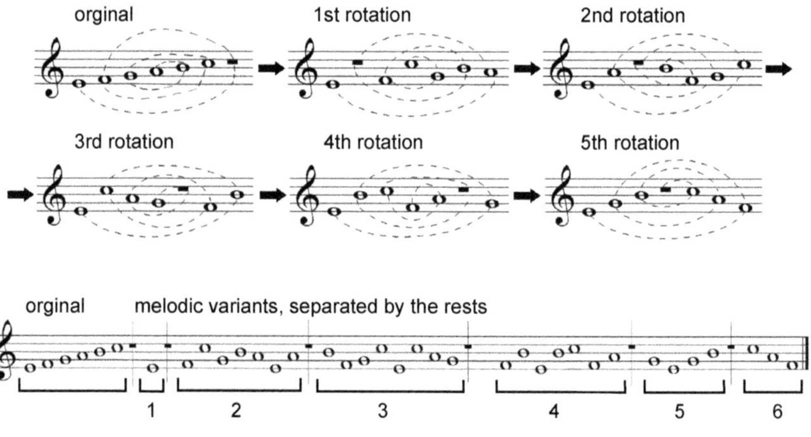

2.4. Personalized Semantics: The Significance of Individual Numbers

Contemporary composers often personify their use of numbers. They use the traditionally defined sacred meanings of numbers and match them, or equate them, with individual messages encoded into their music. Often the meaning of these encoded messages is known only to the composer. For example, Arnold Schönberg (1874–1951) took the meaning of the number 13, which in the numerological tradition is a symbol of fate, and matched it with a biographical detail from his own life – his birthday is 13 September. This is noted in his letter to Webern. He discusses the significance of 1926 for its special connection with the number 13. He writes:

> This is the 52nd year of my life [1874 + 52 = 1926; moreover, 52 = 13 x 4], the 26th year [13 x 2], counting from 1900, which was also the 26th year of my life. And this year – 1926 – I have taken up a new compositional idea.[148]

In this letter Schönberg writes about the idea of the dodecaphonic technique. In May 1926, he completed his Suite, Op. 29, and began his Variations for orchestra, Op. 31. For the first time he used a twelve-tone series in a large-scale orchestral work.

The comportment of number combinations to address personal characteristics is typical of the compositional process of Alban Berg (1885–1935), which is based on numerological intentions. In this composer's music we come across the mystification of the number 23. Often the date on which the composition was completed is indicated as the 23rd day of the month. Berg most likely linked this day with the numerological symbol of his own fate (Gratzer 1993: 9): on 23 July 1908, he experienced his first strong asthma attack and felt the effects of this attack throughout his life. An example can be found in the quartet *The Lyric Suite* (1925–6). Here the composer implicated the motif *a* – *b-flat* – *b* – *f* as the tonal symbolism of his own name, and the name of his beloved, Hanna Fuchs-Robettin (this is especially apparent in the third part of the quartet), and also structurally used the numerological data from his and Hanna's names, the numbers 23 and 10. In this quartet the fatalistic number 23 organized the notes for the metronome marks and in the grouping of measures.

148 The numbers inside square brackets are mine – *R. P.* Quoted in Gratzer 1993: 211.

In Berg's Violin Concerto (1935) we see another symbol that became especially important for the composer in that year – the number 22.[149] This number gives structure to the work's dedication *Dem Andenken eines Engels* (22 letters in German: "To the Memory of an Angel"). In his Concerto, Berg quotes Bach's chorale *Es ist genug* (in the span of 22 measures); in m. 222 we hear the tonal motif, *b-flat – a – c – b*. The date of completion of the composition is written on the last page of the manuscript is 23 July 1935. So, we see the influence of the number 23 in the second part of the Concerto as well – it is made up of 230 measures; the tempo (metronome mark) is noted by the number 69 (23 x 3). In the notes in the margins of the sketch the numbers 10 and 28 are inscribed. Berg either began or ended each important part of the Concerto with a measure that could be divided from 10, 23, or 28.

Structuring individual meaning into symbolic numbers in musical material is illustrated in Lithuanian music by the work of Balys Dvarionas (1904–1972). We see this in his 1972 compositions for piano, the *3 Micropreludes* cycle and *Humoresque.* As his guiding compositional principle, the composer counted the distance between tones by applying an original semantic for the combination of numbers – his phone number and the phone numbers of his family and friends. This is apparent from Dvarionas' manuscripts. On top of the first prelude of *3 Micropreludes* he wrote "23163 (our home phone)." Before his autograph on the second prelude he wrote the numbers "514287", which was the phone number of the composer's daughter pianist Aldona Dvarionaitė. The number combination 58681 of *Humoresque* was not established. However, it is believed that it is also a telephone number, because the piece was composed at the same time as *3 Micropreludes*. The number combinations the composer used became the basis of the construction of the tone motifs. For example, the numerical code of the first prelude can be recognized as the motif *d – e – c – a – e*, which connects with respective numbers once the diatonic scale is transcribed into a line in the traditional method:

c – 1 (an octave higher – 8), *d* – 2, *e* – 3, *f* – 4, *g* – 5, *a* – 6, and *b* – 7

In this manner, the motif *d – e – c – a – e* "intoned" the number combination 23163. Similarly, the intention of the tonal code of the second prelude can be explained – the numerical transcription of the *g – c – f – d – c – b* motif matches with the phone number of his daughter. The origin of the motif of *Humoresque* *f – b-flat – g – b-flat – b-flat* is analogically matched with the number 58681, taking into consideration that this piece was written in B-flat major (see Figure 82).

149 On 22 April 1935 Manon Gropius died, who was the daughter of Berg's close friend.

Figure 82. Dvarionas, Micropreludes *No. 1 & 2 and* Humoresque. *Creation of tone scales according to the phone numbers*

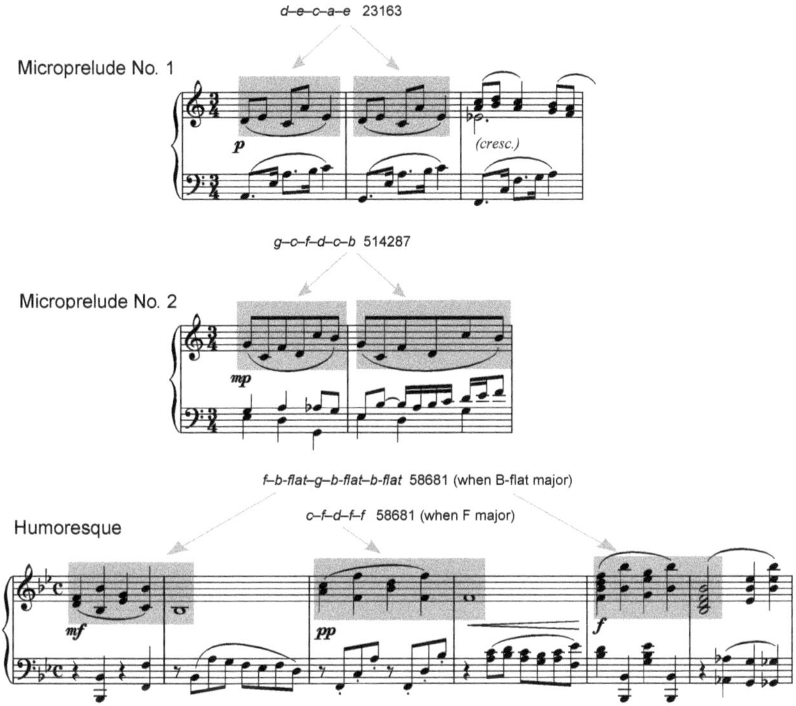

George Crumb. *Black Angels* (1970)

In addition to Crumb's opuses, which were inspired by cosmological themes, the numerologically personified string quartet *Black Angels* (1970) is worthy of attention. In this composition the number appears as a semantic and structural element. The composer's conscious manipulation of numbers is evidenced by the quartet's subtitle "Thirteen Images from the Dark Land", and also a note at the beginning of the composition, a "numerology", as well as the detailed programme of the manifestation of the numbers 7 and 13 in various parts[150] (see Figure 83).

150 The figure of programme notes is based on the composition's score (George Crumb: *Black Angles (Images 1). Electric String Quartet*, New York: C. F. Peters Corporation, 1970).

Figure 83. Crumb, Black Angels. *Semantic and numerological programme by the composer. The connecting arrows display the symmetric arrangement of the quartet parts*

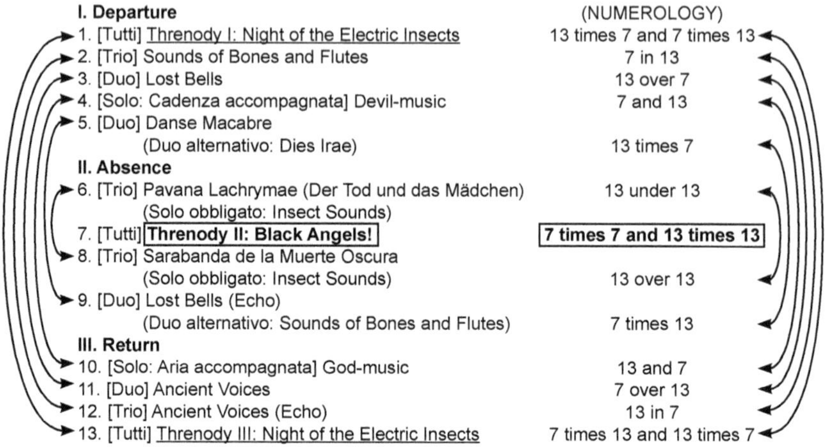

According to Michael Walsh, the narrative of this composition is based on the "battle between God and Satan."[151] We see the analogous dichotomy between the numbers 7 and 13. The number 7 in Christian numerology is associated with divine suffering, with the cross. The devil's number 13 allows the composer to note that in the structure of the quartet there are many musical symbols of this theme, for example *Diabolus in Musica*, *Trillo di Diavolo*.[152] However, the numbers 7 and 13 can be associated with historic events that took place during the Vietnam War, which are important to the composer. At the beginning of *Black Angels*, there is a notation, *in tempore belli, 1970* (Latin – during war time, 1970). At the end of the piece we find "[f]inished on Friday the Thirteenth, March, 1970" (the date matches the end of the Vietnam War).

An analysis of the quartet revealed that the structure of the composition is formed by three numbers, 5, 7, and 13 (the number 5 is also connected with the Friday the Vietnam War ended). For example, the number 13 models the contours of the composition's macroform: the quartet is composed as a cycle of 13 symmetrically resounding parts (1 and 13, 2 and 12, 3 and 11 and so on). The 7th part

151 Based on Walsh's notes on the compact disc of Crumb's work (*Works by George Crumb*, New World Records, 1987, catalogue No. 80357).

152 Crumb's commentary is presented in the CD booklet *Black Angels* (*Kronos Quartet: Black Angels*, Nonesuch, 1990, catalogue No. 79242).

is the center. Each part is marked by the name, which is linked with the "troubled contemporary world." Parts are grouped into three sub-cycles, which recreate, according to Crumb, the three stages of the journey of the soul: 1) departure (i.e., downfall), 2) absence (i.e., destruction of the soul), 3) return (i.e., redemption).[153]

The number seven influences the use of seven tonal instruments (the string quartet: two violins, a viola, and a cello; and three groups of crystal glasses, all tuned differently) or seven percussion instruments (three groups of crystal glasses, two tom-toms, and two maracas). The number five could be associated with five tonal episodes played in the atonal quartet: a significantly modified Dies irae sequence (in parts No. 4 and 5); citation of the melody from Schubert's String Quartet No. 14 *Der Tod und das Mädchen* (part No. 6); the old sarabanda (part No. 8); the aria's melody in cello part harmonized in B major, i.e. five sharps (part No. 10).

The numbers 5, 7, and 13 clearly figure in the vocal part of *Black Angels*. The human voice in general can be heard in five parts (No. 2, 5, 7, 9, and 13). Seven different verbal elements are used in quartet's texts. These are words in six different languages and an abstract phrase used in the second part. In this part two members of the quartet whisper the 13-syllable motif:

ka-to-ko to-ko to-ko to-ko to-ko to-ko

In other parts the verbal text consists of numbers pronounced in six different languages. In part No. 5 musicians count from one to seven in Hungarian. In part No. 7 in Russian, Japanese, German, and Swahili languages musicians say the number 13; and the numbers one through seven are counted in German. In part No. 9 we hear the numbers one through seven in French. In part No. 13 musicians count from one to seven in Japanese twice; the number seven is shouted out twice and the number 13 once. An analysis revealed that the number seven is spoken out a total of seven times in the quartet: once in Hungarian, German and French, and four times in Japanese. The number sequence one through seven is accented five times in the score. The musicians say the number 13 five times in different languages: once in Russian, Swahili and German, and two times in Japanese.

Having analyzed each part of *Black Angels* it was established that the compositional numbers are subject to the logic of the scale and duration. For example, the first part *Threnody I: Night of the Electric Insects* is dominated by rhythmic figures in quintuplets. The small sections are divided into groups of seconds that

153 These titles were taken most likely from the names of the parts of Beethoven's Piano Sonata No. 26 in E-flat major, Op. 81: *Das Lebewohl* (*Les Adieux*), *Abwesenheit* (*L'Absence*), and *Das Wiedersehen* (*Le Retour*).

correspond to three, four or seven second-long durations. The general length of the first part is 91 seconds (7 x 13). This result offers one solution, which could shed light on the composer's handwritten mystical number combination for the first part, "13 times 7 and 7 times 13." The domination of these two numbers is obvious in the structural scheme of the first part (see Figure 84, upper section).

In the second part the parameter of duration is also influenced by the numbers 13 and 7. In the section *b* (marked *b*, b_1, b_2 and b_3; see Figure 84, bottom section) the vocal part is intriguing for its text, the 13-syllable motif *ka-to-ko to-ko to-ko to-ko to-ko to-ko*. Just then, the relationship between the vocal and first violin parts offers a solution to this part's numerological riddle "7 in 13." Because at once, we hear the first violin playing seven intervals of sevenths and a 13-syllable vocalization. One other means of deciphering the composer's riddle was seen in section c. Here the 13-tone fragment is divided into three parts of 3, 7, and 3 tones. The seven sixteenth-note motif is framed between the thirty-second notes. That same rhythmic formula does not change in all *c* sections, and only the interval structure changes (see Figure 84, bottom section).

When analysing the other parts of this quartet[154] I noticed that the influence of Crumb's chosen numbers can be seen on various levels. For example, the symmetrical form of the third part is made up of seven compositional blocks. The number seven influences the quantity of tones in separate blocks (see Figure 85, upper section). The symmetrical aspects of the form are repeated in the 11th part, which is made up of blocks of seven tones. The symmetry is unique in the arrangement of the interior of the musical material as well.

In the fourth part, *Vox diaboli*, the devil's voice is symbolized by conceptual numbers 7 and 13, which influence the number of tones and their duration while using the sequence of Dies irae. The continuance of the death themes in the fifth part *Dance macabre* takes place in 25 measures (5 x 5). At the end the composer indicates that the sixth part begins after a 13 second rest (see Figure 85, bottom section). The sixth part is composed from sections of 13 seconds each; in total, thirteen tones play in these sections (bearing in mind the grace notes). That same logic is repeated in the construction of the eighth part. In the center of the cycle, in the seventh part, in a variety of languages, we hear the numbers one through seven called out; the number thirteen is called out separately, and so on. The ninth part is concluded with an analogical indication, as in the end of the fifth part. Its structure is based on number constructions that echo the structure of the third part.

154 A detailed analysis of Crumb's *Black Angels* was part of this author's Master's thesis at the Lithuanian Academy of Music and Theatre (see Povilionienė 2003) and was published in the article "Technical and Symbolic Numerical Construction in George Crumb's *Black Angels* and György Ligeti's *Etudes*" (see Povilionienė 2004).

Figure 84. Crumb, Black Angels. *Above: Organization of rhythmic figures in quintuplets in the first part; below: structural diagram of the second part*

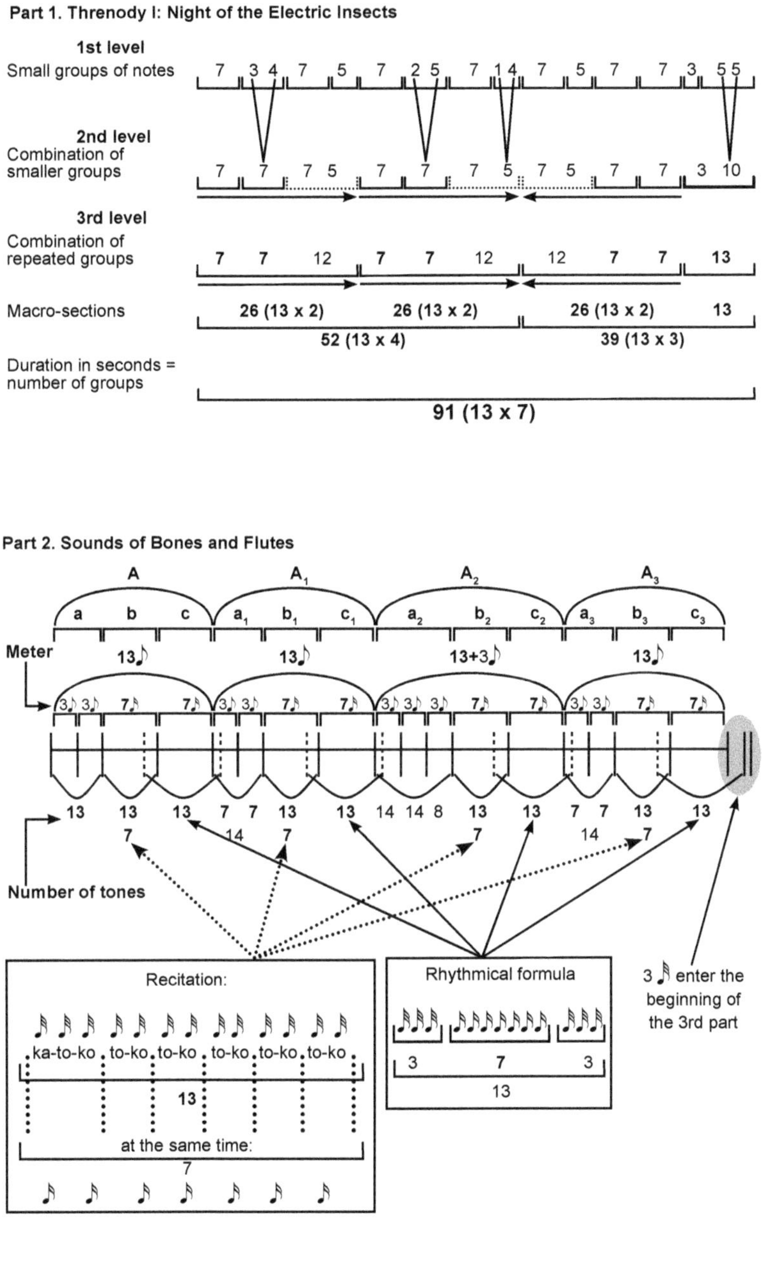

Figure 85. Crumb, Black Angels. *Manifestation of numeral 7 and 13 in the structural diagrams of the third (above) and the fifth (below) parts*

Part 3. Lost Bells

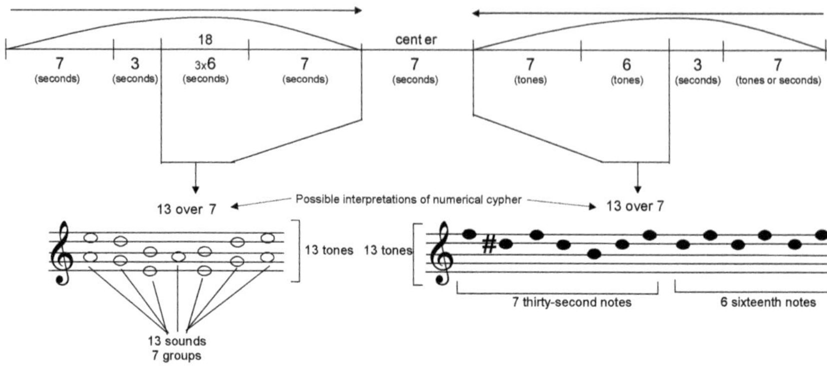

Part 5. Danse Macabre (Duo alternativo: Dies Irae)

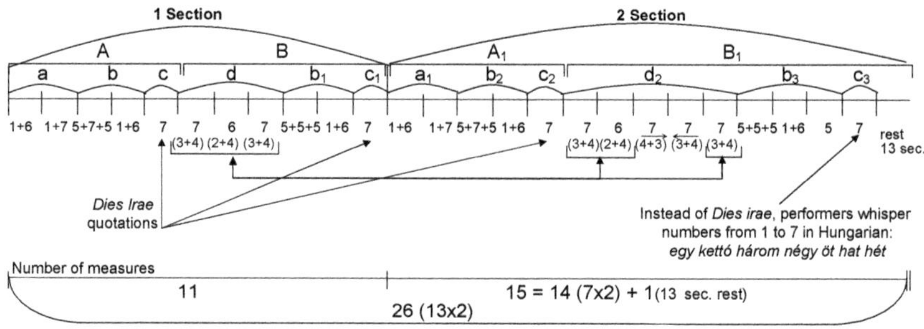

John Cage. *Ryoanji* (1983)

Cage's eight-part composition *Ryoanji* for solo instruments and voice was composed in 1983 in Seattle. It is possible to describe this piece as the semantic expression of the Ryoanji rock garden in Japan.[155] With this musical composition Cage complemented his cycle of 15 computer-generated drawings and engravings, titled *Where R = Ryoanji.*

The harmony of the Japanese garden is constructed from 15 different size stones, which are tossed out in piles within the waves of sand. It is interesting that at one time, if one were to gaze at the garden from any corner, it is possible to see only 14 stones. The area of the garden is 2340 square feet (30 x 78). The stones are arranged according to five groups. When gazing from the left to the right their quantity in separate piles matches the number sequence 5–2–3–2–3.

The visual interpretation of the garden can be split into two macrogroups of stones. One macrogroup is made up of two piles of stones on the left part of the garden (a composition of 5 and 2 stones, making a total of 7 stones). The second macrogroup is made up of the other three stone piles on the right (this group consists of 3, 2, and 3 stones, making a total of 8 stones; see Figure 86).

The allegory with the Ryoanji garden elements is obvious in the macrostructure of Cage's musical composition. The cycle *Ryoanji* is made up of eight independent parts (correlating with the eight gardens in the temple). Each part of the cycle must be performed in a different part of the concert hall (as though to create an allusion to the scattering of the stones or gardens). The central part is performed by percussion. Its sound, according to Cage, echoes back the texture of the raked sand.[156] This part is made only of pulsating quarter notes and quarter-note rests. The composer probably used the numerical peculiarities of the musical composition's prototype, striving to repeat the parameters of the rock garden in the plan of the score. For example, the 78-foot length of the garden just about repeats the piece's length of 77 measures. An analysis revealed certain commonalities in the composition:

---

155 The most famous contemporary rock garden can be found in Kyoto in the Ryoanji Temple, on the southern half of the temple (in 1994 the temple was listed as a UNESCO World Heritage Site). It is believed that the rock garden was established in 1488.

156 According to the notes by John Cage in the score: *Ryoanji*, New York: Henmar Press, 1983, p. 2.

Figure 86. *Cage,* Ryoanji. *Stone composition in Ryoanji garden as the allegorical source for the arrangement of rests in the percussion part*

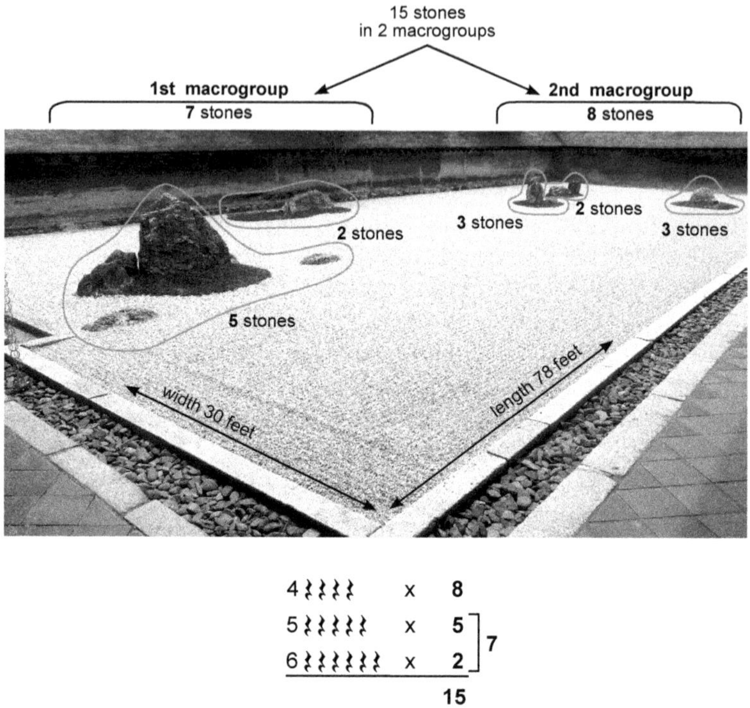

1) each measure in the colorful musical fabric begins with the beat of a quarter note (not one measure begins on a rest);
2) all the measures (except for m. 75)[157] has a different rhythmic formula that is made up of five quarter notes and a changing combination of the number of rests. This is the first possible allegory with the five piles of stones tossed out in the garden and one of the piles that is made up of five stones;
3) the one and two quarter-note sequences dominate; a sequence of three quarter-notes is heard only three times.

---

157 The number 75 can also be found in the interpretation of the measurements of the Ryoanji rock garden. According to the Japanese system of measurement, the area of the garden is equal to 75 tsubo (a measurement used in Japan to measure units of area).

In the analysis of the compositional logic of the rests I noticed a visual and constructive coincidence with the numerical codes of the rock garden. Though Cage ordinarily manipulated with groups of one, two, or three quarter-note rests, he inserted single blocks of 4, 5, or 6 quarter-note rests that were tossed around chaotically. When I counted the number of times they appear in the score and where they appear, I found a correlation: Cage brings in the groups of four quarter-note rests eight times; he brings in the groups of five quarter-note rests five times; and he brings in the groups of six quarter-note rests two times (see Figure 86).

All the numbers inspired me to think that there is a direct link between the constructional elements of the Japanese rock garden and Cage's composition. The general number of appearance of 4, 5, or 6 quarter-note rests – 15 – matches with the number of stones scattered in the rock garden. The numbers of the individual groups, 8, 5, and 2, correlate with the interpretation of the visual composition of the rock garden – the two macrocompositions of stones with seven (5 + 2) and eight stones each. The similarities of the visual musical score with the configuration of the layout of the stones allow one to believe that the chaotic placement of the rests in the score resounds with the garden chart, 5–2 (7) and 3–2–3 (8). It could be stated that the actual rock garden was not only the first idea for the composer, but also the organizing means of forming the musical composition (compare Figures 86 & 87).

Figure 87. Cage, Ryoanji*, percussion part. Diagram of the arrangement of rest groups in the score*

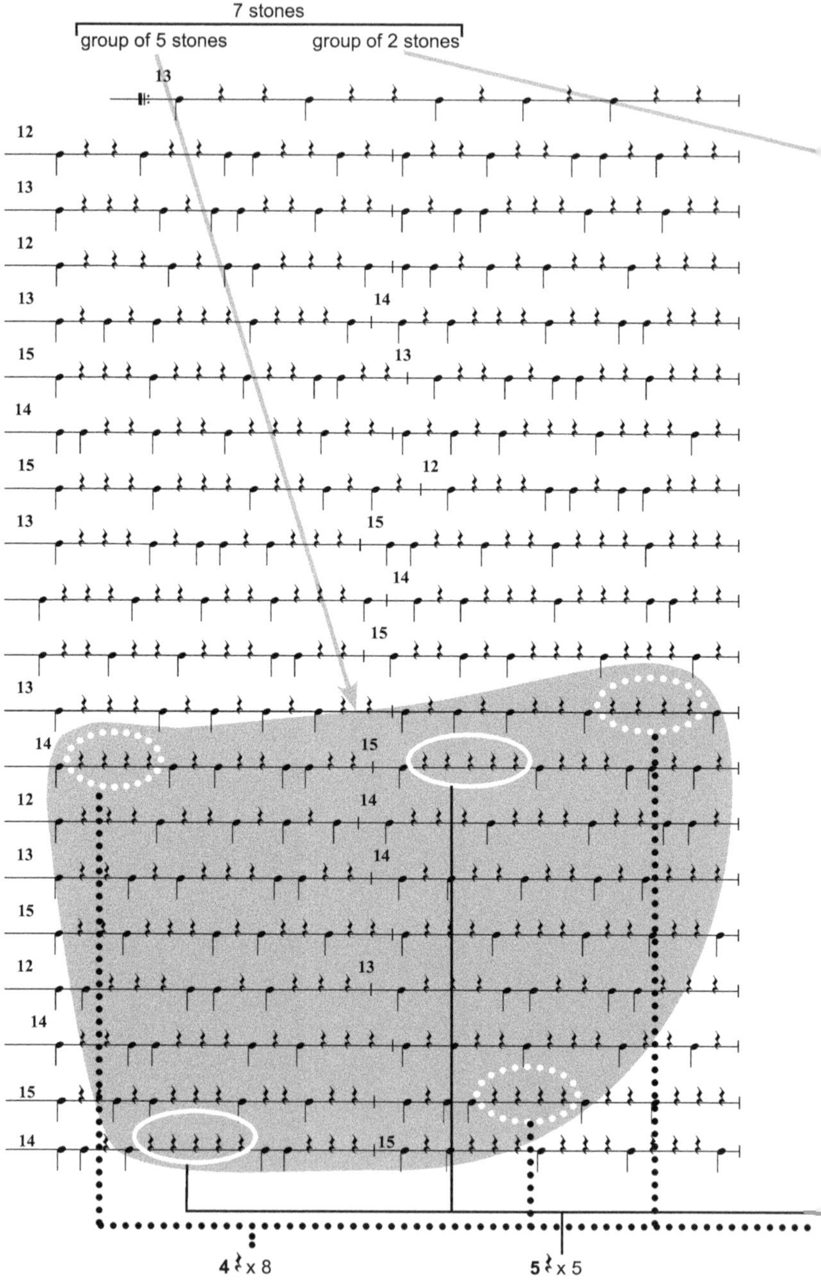

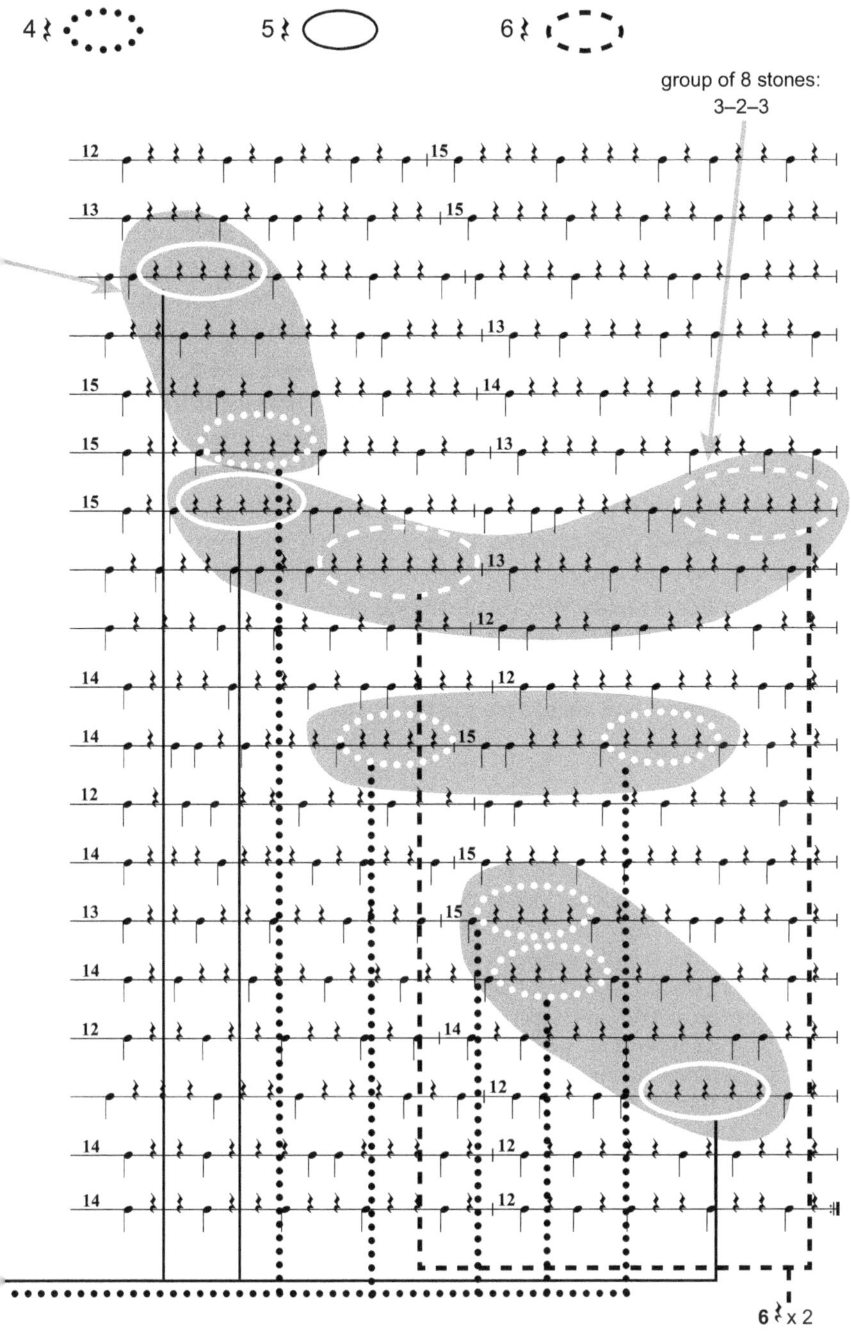
4
5
6
group of 8 stones:
3–2–3
6 x 2

Mikalojus Konstantinas Čiurlionis. Fugue in B-flat, Op. 34/VL 345 (*1908–9*)

Čiurlionis' Piano Fugue in B-flat (1908–9) can be interpreted as the musical expression of the composer's autograph.[158] This premise is based on examples of Čiurlionis' paintings, his drawings, and his correspondence (e.g., his letters to his brother, Povilas, and his wife, Sofija Čiurlionienė-Kymantaitė) as well as numerological semantics.

After I analyzed the structure of the subject of the fugue, in addition to other details of the composition's construction, I could see certain tendencies, which allowed me to surmise that the piece for piano contained not only visual and emotional connections with the idea of the beginning of the opera *Jūratė* (e.g., the semantics of the depths of the sea), but also a rational, and preconceived, perhaps even calculated, method of manipulating the musical material. Most likely, Čiurlionis' conscious manipulation of two numbers, 11 and 22, played a role. Additionally, the structure of the subject tone-scale can be interpreted numerologically along with the organization of duration or with the peculiarities of the rhythmic picture. For example, in the two music score versions (edited by Vytautas Landsbergis and Jadvyga Čiurlionytė) the subject is presented as a structure of 21 tones. Antanas Venckus also based his interpretation on a 21 tone span (Venckus 2000: 175–253), although in Rimantas Janeliauskas' opinion the length of the subject is defined by 25 tones (Janeliauskas 2010: 335). We see that there is no consesnsus, how many tones make up the subject of the fugue. This

158 The fact that in his music Čiurlionis used the alphabet and original cryptographic symbols has been confirmed by several researchers of his work (Vytautas Landsbergis, Jadvyga Čiurlionytė, Gražina Daunoravičienė, Darius Kučinskas and others). There is also evidence that was left behind by the composer himself. In his manuscripts from 1906 the composer wrote a code to decipher equivalents between musical tones and the corresponding alphabet (manuscript in M. K. Čiurlionis National Museum of Art, Čm–21, page 260). There is also his musical signature, which shows how in a manuscript a potential idea for a composition is born (manuscript in M. K. Čiurlionis National Museum of Art, Čm–6, page 411).
In my study "Search of Čiurlionis Signature and Dedication. Fugue in B-flat: Structural and Semantic Analysis" (see Povilionienė 2013) I have published a research of Čiurlionis' semantics in his painting, his sketches, his manuscripts, and in his print *Kompozicija* (Composition, 1909), which reveals an ornamental secret code that is subtly dedicated to his wife, Sofija. Additionally, the article includes a comprehensive historical and methodological analysis of the graphics and numerology of the Fugue in B-flat.

led me to one more interpretation – of the 22-tone subject. Such a hypothesis allows one to raise a few other observations that show the use of two related numbers in the texture of the music, 11 and 22:

1) the fugue subject is made up of 11 different tones (instead of all 12 chromatic semitones):
   b-flat – b – c – d-flat – d – e-flat – e – f – g-flat – (the missing tone *g*) – *a-flat – a*;
2) the subject is displayed in the fugue a total of 16 times; however, only 11 times it is played in its complete form. In all the other instances the theme is incomplete (of 11 or 18 tones);
3) if the subject has 22 tones, then by analyzing its duration it is apparent that it is made up of 29 eighth notes. And the Golden Ratio appears before the 18th eighth note (that is 29 eighth notes x 0.618 = 17.922). When I analyzed the place where it appears in the subject, I noticed that the Golden Ratio matched perfectly with the climax of the subject. It is the highest tone *f*, which is the eleventh out of a total of 22 tones in the subject (see Figure 88);
4) the manipulation of number 11 is obvious in the construction of the subject's rhythmic pattern, which is divided into two sections, each of 11 eighth notes. In the first section, the group of 11 tones features an inaccurate rhythmic symmetry, which is framed in quarter notes, with a general length of 11 eighth notes. In the second section the group of 11 eighth notes moves evenly. This same partitioning of the subject's rhythm into two parts was noted by Lithuanian musicologist Venckus (Venckus 2000: 178). However, he does not talk about the symmetry in the first half of the subject;
5) the logic of the number 22 can be seen in the fugue's preamble (in the five-measure introduction). Later, Čiurlionis got rid of this part; the introduction only remains in the first manuscript from 1908–9 (stored in M. K. Čiurlionis National Museum of Art, Čm–52, pages 20–8). The introduction is made up of the descent of the three-tone motif *f – a – c-sharp* in the part for the right hand and contains a total of 22 tones (see Figure 89).

The obvious tendencies of the numbers 22 and 11 raises the question – are they still a compositional construction? Or are they accompanied by a semantic subtext? Within the context of hypotheses about the use of numerological codes in the work of Čiurlionis these numbers take on a personalized subtext. Firstly, one can see a connection between the artist's date of birth written in

Figure 88. Čiurlionis, Piano Fugue in B-flat. Diagram of the Golden Ratio of the fugue subject and the manifestation of number 11 in the arrangement of rhythm

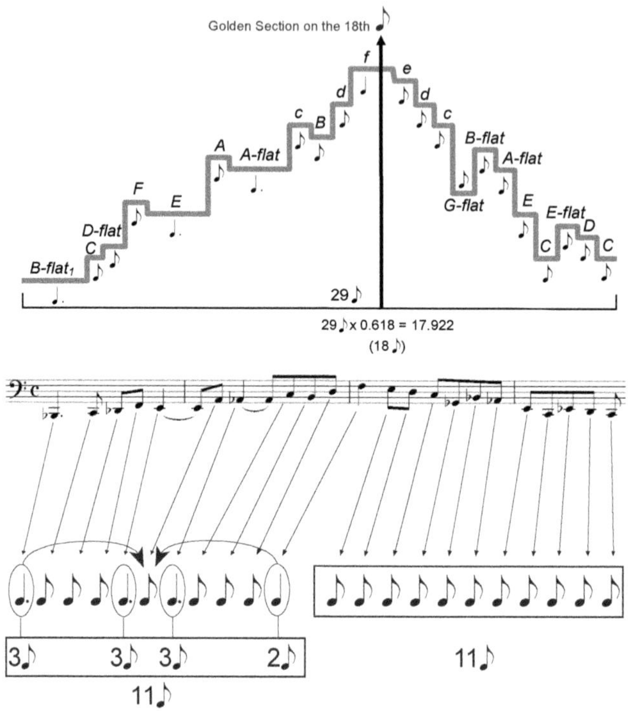

Figure 89. Čiurlionis, Piano Fugue in B-flat. Fragment of the preamble in the first manuscript version

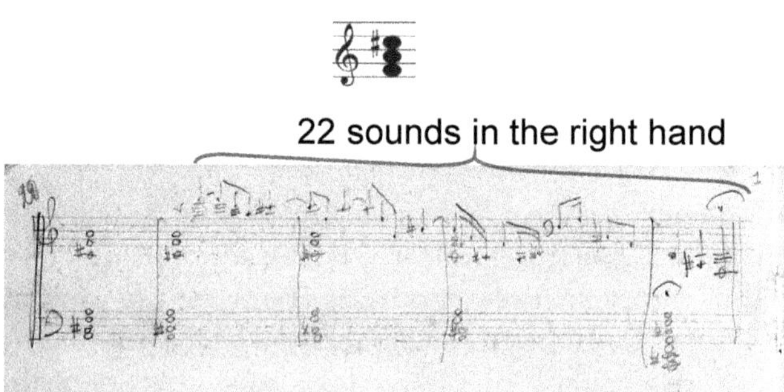

Figure 90. Different combinations of 11 letters in Čiurlionis' signature

the new style, 1875, September 22.[159] Čiurlionis was born 1875, September 10 according to the old style, which is how his date of birth is notated in the church records. However, in the first Lithuanian newspaper, *Vilniaus žinios* (Vilnius News) starting from 1905, on the title page the date is printed in both styles. Both Čiurlionis and his wife, Sofija, were regular readers of this newspaper, as well as contributors. Additionally, the composer had to contend with his birth date being written in different ways during his lifetime: in Druskininkiai, St. Petersburg, Warsaw and Vilnius, he lived according to the old style. He lived according to the new style for one year of his studies, 1901–2, which he spent in Leipzig. Later, in 1906, he visited Western Europe (Vienna, Munich, Prague, Dresden, Nuremberg). This dual manner of recording dates and their confusion is confirmed by Čiurlionis' friend Jonas Zaluska in his 1908 December 4 letter to the composer. This letter is connected with setting down Čiurlionis' wedding date. Jonas asks: "Write and tell me whether you mean the New Year in the old style or in the new" (Čiurlionis 2011: 256). It is obvious that Čiurlionis knew the doubling of his date of birth (September 10 and 22). Yet, whether he himself brought meaning to this doubling is not clearly known. There is no direct evidence in his correspondence.

Secondly, the argument that these numbers were not chosen coincidentally is raised by the connection with Čiurlionis' signature in both Lithuanian, *M K ČURLIONIS* (also, *M K ČURLIANIS*), and Polish, *M K CZURLANIS*. In all instances his signature is made up of 11 letters (see Figure 90). These observations allow one to argue that Čiurlionis, having surmised his intriguing signature and its connection with the new manner of writing dates (11 is the largest proper divisor of the number 22), could have adapted it in his music. The Fugue in B-flat is his last composition for piano and is considered symbolically significant as Čiurlionis' musical signature.

---

159 In the 19th century, after Lithuania was annexed by the Russian Empire, the Gregorian Calendar (still called the new style) was reverted to the Julian Calendar. The latter was valid in Lithuania up until the end of World War I. Then, the style for writing dates returns to the new style. In the 19th century, the Julian Calendar differed from the new style by 12 days and in the 20th century by 13 days.

Part 3
Innovations of Mathematical Techniques in 20th and 21st Century Music

In the diverse practice of creating music in the 20th–21st centuries, the innovations of mathematical processes related to the application of new mathematical theories, which considerably broadened the space of creative possibilities and the problem field, became established on equal terms alongside traditionally determined phenomena of a mathematical nature. The principles taken over from more advanced mathematics, fast developing spheres of information technologies, began to be transformed into a practice of musical composition in the middle of the 20th century. This is a trend of writing music in complicated algorithms that has become possible in the age of the computer, allowing the creation of geometrical, graphic algorithms as a prototype of a composition and ways of a schematized expression of a musical composition, as well as the explication of complicated mathematical formulas or models, mathematical theories (fractals, chaos, groups, probabilities and others), practice of scholastic music, etc.

The abstracted mathematical basis is also peculiar to analytical approaches of contemporary music. Here, numerical proportions based on the principles of set theory[160] are applied, mathematical interpretations of musical structures are being formed, cases of transforming music into geometrical models and vice versa, mathematical expressions applied in the analysis of sound pitches and tuning are studied. The innovative phenomenon of rendering contemporary music mathematically inspired musicological literature of new kind that investigates algorithms and their procedures in the musical tissue, application of recursive models (as well as fractals), and chaos theory or probability theory in creating computer music.

It can be said that the main novelty in the composing of music in the 20th–21st centuries is the models used for numericalization of sonic structures taken over from formal grammar, geometry of fractals, chaos and other modern mathematical theories. Therefore, on the basis of rather different practical cases of the implementation of mathematical innovations in music and research of the sources that investigate that sphere, the examples of contemporary music that are discussed further are systematized according to unifying technologies. They are as follows:

- graphics of music that has been rendered mathematically. Here attention is drawn to the graphic structures of pitches, prototypic models of graphics, which are chosen as a tool for both composing and later commenting on a composition. Numerically writing down of the musical score, the phenomenon

160 According to the Allen Forte's set theory, the structure of music composition can be reduced to sets of pitches and their intervallic relationships. Composer Milton Babbitt is considered to be the creator of set theory analysis.

of original numerical notation, the application of L-system models, as well as principles of structural linguistics in a musical composition, etc. can be attributed to this trend;

- the establishment of the conception of algorithmic composing of music and application of algorithmic processes in the sphere of composing contemporary music, which makes an ever wider use of computer device and which has opened immense possibilities for generating musical ideas and their implementation;
- musical representations of the theory of fractals, analogues of fractal geometry in the structures of musical compositions (pseudo-fractals, fractal images, the fractal principle as a means of composing) and a spread of other theories of contemporary mathematics (practice of stochastic music, Markov chains, the impact of the theories of chaos, probabilities, transformations, and groups on the composing of music).

1. The Mathematized Musical Graph

The structuring of musical time and space according to exact dimensions creates the opportunity to transcribe tonal material into a graph and to recreate the entirety of the composition into specific geometric figures. Such a visual tool enables one to see the tendencies of hidden relationships of tones in a score and to give them a mathematical grounding. In order to construct a graphic of a musical texture, an especially appropriate device is a two-dimensional plane for *x-y*. Usually, the *x* and *y* axes are filled with the data of two musical compositional parameters – pitch and duration. For example, the pitch and duration of the first, highest, lowest, and last tones can be put down as certain mathematical units or coordinates, which can be connected into the appropriate curve. Such an approach to the visual analysis of a musical composition is applied by Alexander R. Brinkman and Martha R. Mesiti (see Brinkman & Mesiti 1991), Roy Howat (see Howat 1983), Pozzi Escot (see Escot 1999) and others.

The parameters of these two fundamental musical elements (pitch and duration) as visualized on the *x-y* plane can be implemented to analyze musical compositions from a variety of epochs. As Escot argues, it is "another way of notating the music which vividly conveys a different perspective" (Escot 1999: 6). This kind of approach may reveal that Hildegard von Bingen's antiphons and Györgi Ligeti's compositions have the same musical architectonics, which is based on the same dimensions.

Moreover, computer technologies offer the possibility to graphically analyze much more complex tonal elements than heretofore research applied. The computer allows researchers to apply the acoustic methods, assemble a photograph of the tonal spectrum, a sonogram. This was done in a practical way, for example, by Alfred Lorenz. He used diagrams and curves to analyze the modulation processes of Richard Wagner's *Der Ring des Nibelungen* (Brinkman & Mesiti 1991: 1–3).

1.1. A Geometric Prototype as an Algorithm for Musical Composition

A visual model may not only be the respective result of the analysis of a musical composition. It may also serve as a first impression, idea, or a structural algorithm. For example, two geometrical curves influence the selection of pitch in Tom Johnson's piece *Two curves* for player piano (1998; see Johnson *Editions 75*). The seven-sided figure determined the choice of seven instruments in Alain Louvier's *Heptagone* for piano, harp, flute, clarinet, violin, viola and cello (Heptagon, 2003). An example of the relationship between geometry and music could be the cycle *Heksagoni* (Hexagons, 1975–8)[161] by the electronic music composer Srđan Hofman (born 1944). In the part for cello, *Monodrama,* the number six (most likely as an analogy for a geometric figure) influences the structure of the scale. In this part one hears an incomplete scale of harmonic major because of the omitted fourth degree, for example, *c – d – e – g – a-flat – b*. The metronome marks written in the piece, 60, 72, 120, 132, 264, and 360, are also linked with the number 6 as the common divisor.

I will provide a few more examples of the practical application of geometric figures in the process of musical composition. For example, one might consider Dutch composer Jan Rokus van Roosendael's (1960–2005) composition *Rotations* for flute (1988). In the structural sections of this composition one can identify the composer's sketches of geometric models (a variety of four-sided figures), which act as the organizational order of different musical groups. Another example would be John Adams' composition for piano *China Gates* (1977). In this piece the parameter of the duration repeats the contours of a curve drawn into the first page of the score.

Jan Rokus van Roosendael. *Rotations* (1988)

Roosendael's compositional style is known for the repetition of rhythmic models and the strict formation of rhythmic parameters. This can be seen in the manner in which the composition *Rotations* was put together. In the beginning of the score the composer draw a geometric graph with a circle in which the twelve compositional segments are laid out. This figure-prototype formed the musical composition and influenced the combination of different segments (see Figures 91 & 92).

161 The cycle is made up of four instrumental pieces: the first – *Farsa* (Farce) for piano trio (violin, cello and piano), the second – *Monodrama* for cello solo, the third – *Pastorala* (Pastoral) for violin solo and the fourth – a musical scene with theatrical elements, *Ritual* for instrumental ensemble and six groups of women.

Figure 91. Roosendael, Rotations. *Geometrical prototype (the diagram reproduced from Jan Rokus van Roosendael's score:* Rotations, *Amsterdam: Donemus, cop., 1988)*

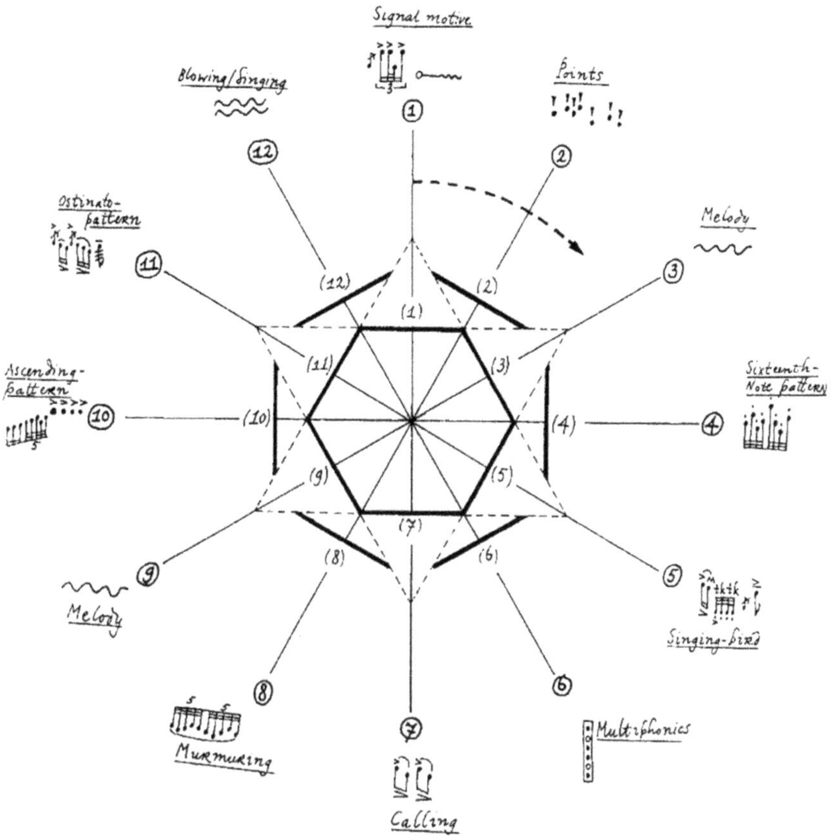

An analysis revealed that the beginnings of the 15 sections of the piece (which are marked with different letters, A, B, C…) have several invariant properties. Firstly, all the sections (with the exception of the 11th section K) were organized using a repetition of segment groups 1–2–3 and 7–8–9. Secondly, when I made a graph of the score to analyze the segments that enter into the composition (Figure 93), it became clear that the first ten parts were composed by consistently adding or taking away two segments. A tendency towards symmetry reflected the choice of the number of segments. In the other sections the linearity was imperfect – some segments were skipped, or their order in their lines was changed around.

Figure 92. Roosendael, Rotations. *Diagram of segment groups on the first page of the score*

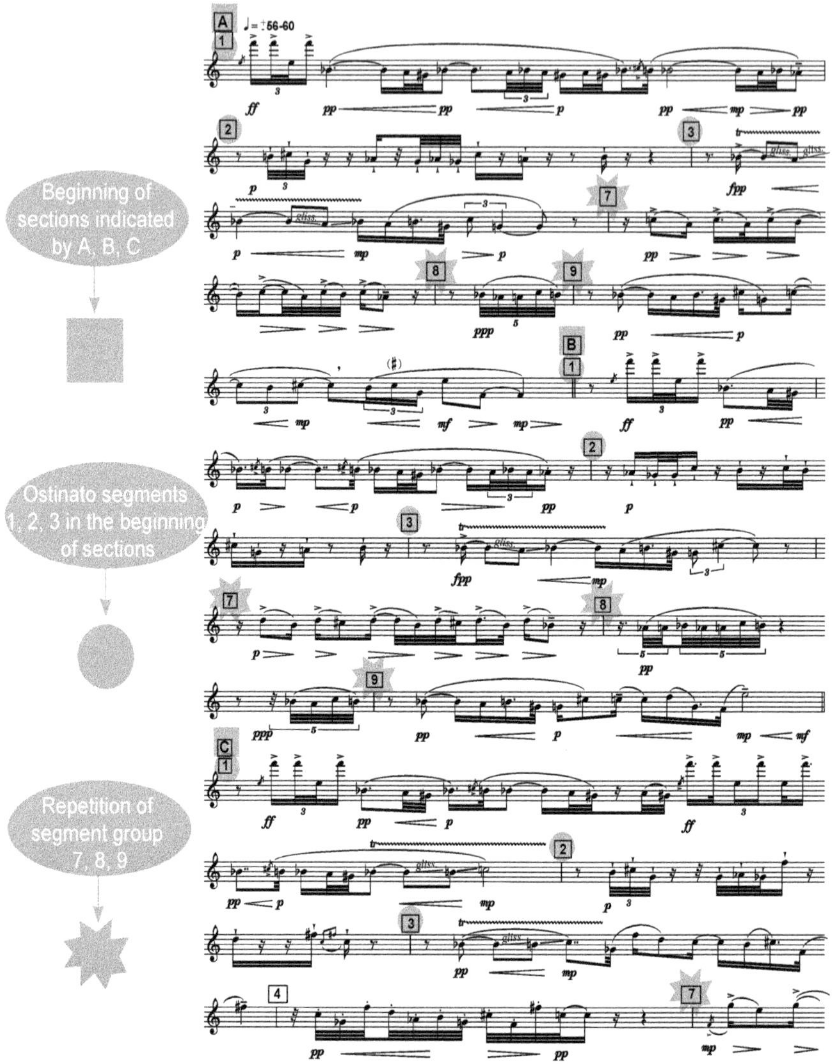

Figure 93. Roosendael, Rotations. *Diagram, showing the symmetry of adding or taking away the segments*

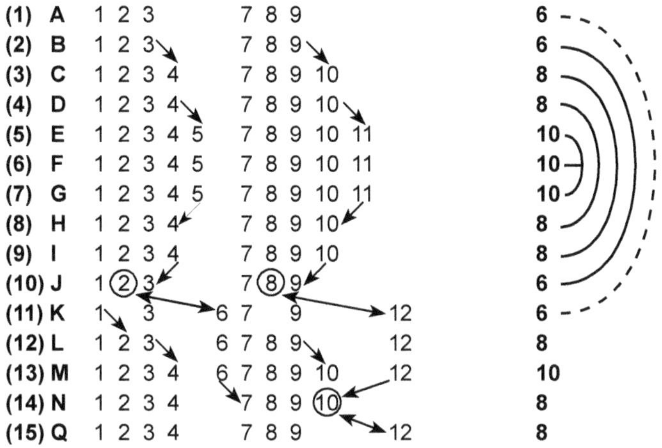

Later, I tried to establish connections in the musical material with elements of the scheme, i.e. the complex of geometric figures in the center of the graph presented in Figure 91. The lines were connected at each section's small group's segments along the edges (for example, in section A there are two segment groups, 1–2–3 and 7–8–9, which on a linear level are connected with the segments on the edge, 1–3–7–9). As it turned out, the structure of the parts of the piece was dictated by four geometric figures – the rectangle, the square, the rectangle with two corners cut off and the trapezoid (see Figure 94). For example, the graph of musical segments of section A has the form of the geometric figure of a rotated rectangle. This is also typical of the structure of the B and J sections. The shape of another rotated rectangle can be found in the E, F, G and M sections. The figure of the square became the graphic expression of five sections with the following sequence of musical segments: 1–2–3–4–7–8–9–10. The square's shape is typical of the L section as well. The contours of the last part (Q), 4–7–9–12, were formed into the trapezoid form (segments 12–1–2–3–4–7–8–9). It was also established that the form of an imperfect rectangle was included into the composition. This was in the K section, which made up the figure of the rectangle with the corners cut off. When I laid out the derived geometric figures in a line, in their sequence I saw a tendency towards the repetition of the shapes, which influenced the piece's structure from two compositionally symmetrical blocks (see Figure 95).

Figure 94. Roosendael, Rotations. *Four geometrical models, used for the organization of sound material (rectangle, square, rectangle with the corners cut off and trapezoid)*

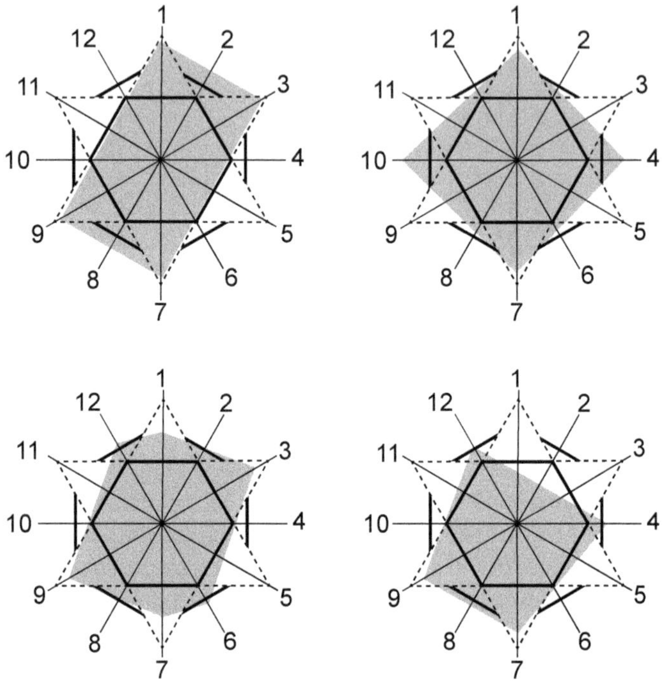

Figure 95. Roosendael, Rotations. *Musical structure as symmetrical arrangement of geometrical shapes*

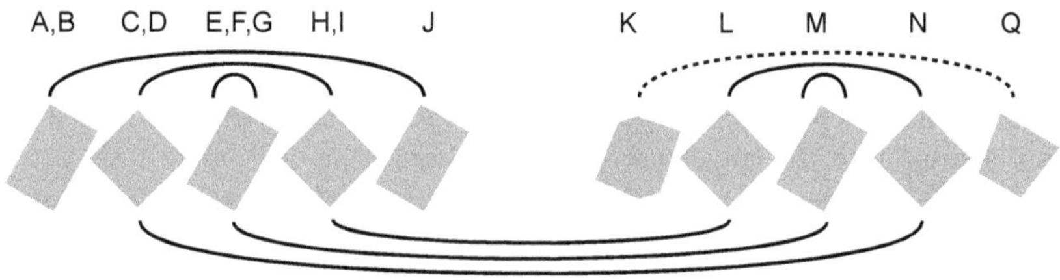

John Adams. *China Gates* (1977)

One of the first American composers to consciously turn away from the aesthetic of postwar European and American academic avant-garde and to work within the realm of minimalist music was John Adams (born 1947). Typical of his music is a hypnotic, trance-inducing pulsation, which slowly opens up into a harmony. Adams' first minimalist experiments were *Phrygian Gates* (1977) and *Shaker Loops* (1978). At that time, he also composed his piece *China Gates* (1977). This composition for piano represents the use of numerical operations in the compositional process. The general structure of the piece is explained as the realization of a graphic curve. On the first page of the score Adams himself drew in a curve and beside it wrote the concept "gating." That indicated that every turn of the graphic curve had a musical equivalent of a change in the musical phase, a transition from one mode to another.

The analysis of *China Gates* showed that the geometric prototype determined the musical parameter of duration. The length of the horizontal line influenced the number of quarter-notes in different phases. The turns of the curve in the music revealed themselves as a modal change (a perpetual exchange of five flats and sharps) and the entry of a new repetitive tonal group. Therefore, according to the curve's rotation ("gating") the exchange of structural blocks in music was accelerated or slowed down. For example, in the first section we hear four pairs of structural blocks, which have the tendency to become shorter by 15 quarter-notes (60–45–30–15; which is the arithmetic ratio 4–3–2–1): after the first two phases, which have a length of 60 quarter-notes, we hear a third and fourth phases of 45 quarter notes, then the fifth and sixth phases are each 30 quarter notes, and the final phase pair is of each 15 quarter notes (see Figures 96 & 97). An analogical process of changing phases ("gating"), only growing rarer, is reflected in the third section of *China Gates*. Moreover, a visual connection of the first and third sections establishes a symmetrical construction (see Figure 98). The center section stands out for its very dense phase exchange (just like the curve's turn becomes more active in the graphic centre). The number of its measures is directly dependent on the number sequence, which reveals the relationship of the geometric proportions, 2–4–8–16, and so on (see Figure 98).

Figure 96. Adams, China Gates. *Interchange of musical phases in mm. 1–42*

Figure 97. Adams, China Gates. *Diagram of the first section according to the duration of phases*

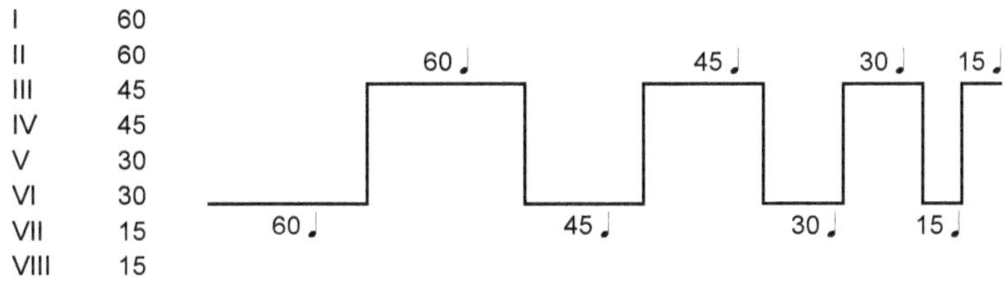

Figure 98. Adams, China Gates. *Graphical arrangement of the whole musical piece*

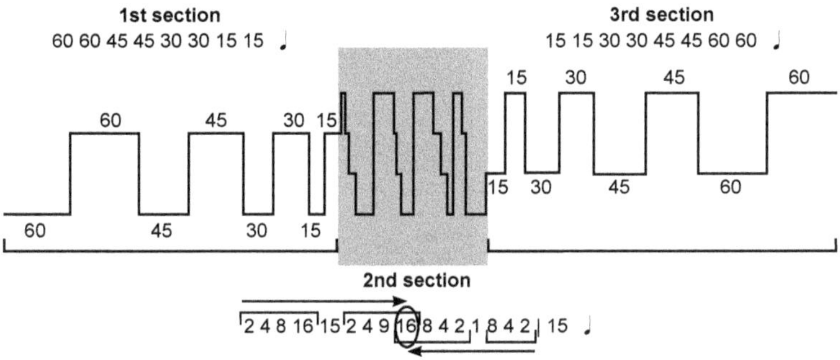

1.2. L-system Formalities in Music

The graphic prototype of a musical composition can be borrowed from an idea that comes from a different branch of scholarly pursuit. A good example would be the application of the principles of formal grammar (the L-system). This is a direction in structuralism that became widespread in the 20th century, and which influenced a number of humanities – philosophy, psychology, linguistics, anthropology, and music. This scholarly branch originated when the early linguist Ferdinand de Saussure (1857–1913) raised the idea that "the main relationships between units of language can be expressed through mathematical formulas" (Gumauskaitė 2000: 65).[162]

162 This thought by de Saussure predicted one of the directions of structural linguistics – the generative grammar theory. In the field of psychology, structuralism appeared as the *Gestaltpsychologie* theory, which opens up for a discussion on constructive elements of the brain and the basis of conscious data, and which argues that the human brain not only reflects the real world in a profound manner, but organizes and structures behavior and understanding.

Figure 99. Creation of musical counterpoint of two voices according to a Gosper curve and its rotation

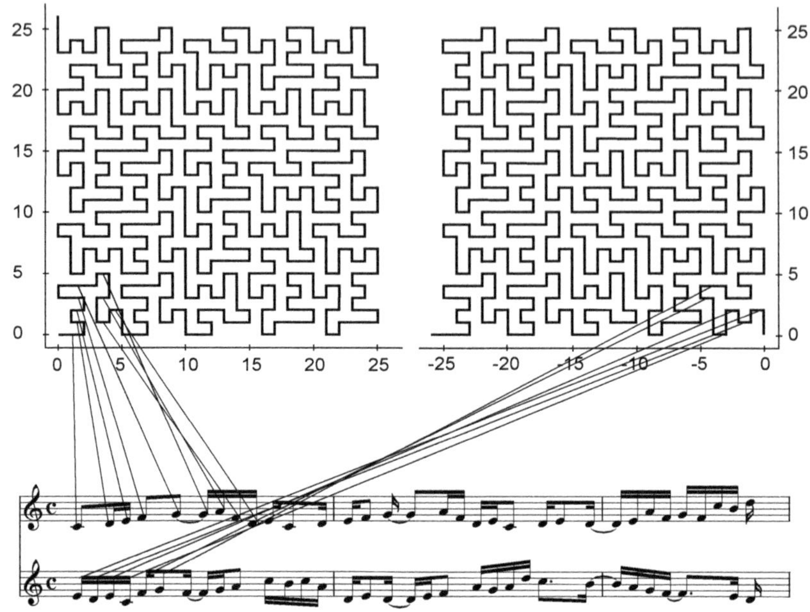

The proliferation of the principles of structural linguistics are obvious in contemporary computer music. According to Hanna Järveläinen, the algorhythmic L-system (a Lindenmayer system[163]) formulas are based on the rules of formal grammar (Järveläinen 2000: 7). These formulas can be applied to the process of generating music on the computer and transforming the algorithms into tonal structures. Music can be composed on the computer by applying the grammatical paradigms of the American linguist Noam Chomsky (born 1928) as well. One method of composing music by using the L-system curves was suggested by Przemysław Prusinkiewicz, who adapted the *x-y* plane principle. The horizontal line of the curve is compared with the duration of musical tones, while the vertical line is equivalent to pitch (Prusinkiewicz 1986: 456; also see Mason & Saffle 1994: 32). For example, the Gosper curve can create a musical counterpoint of two voices, and from that same curve it is possible to find four rotations and a total of eight musical motifs (see Figures 99 & 100).

163 The L-system is named after the Hungarian biologist Aristid Lindenmayer (1925–1989). This theory explains models for the growth of living organisms, which are typical of recursion.

Figure 100. Arrangement of musical motifs according to the rotations of a Gosper curve

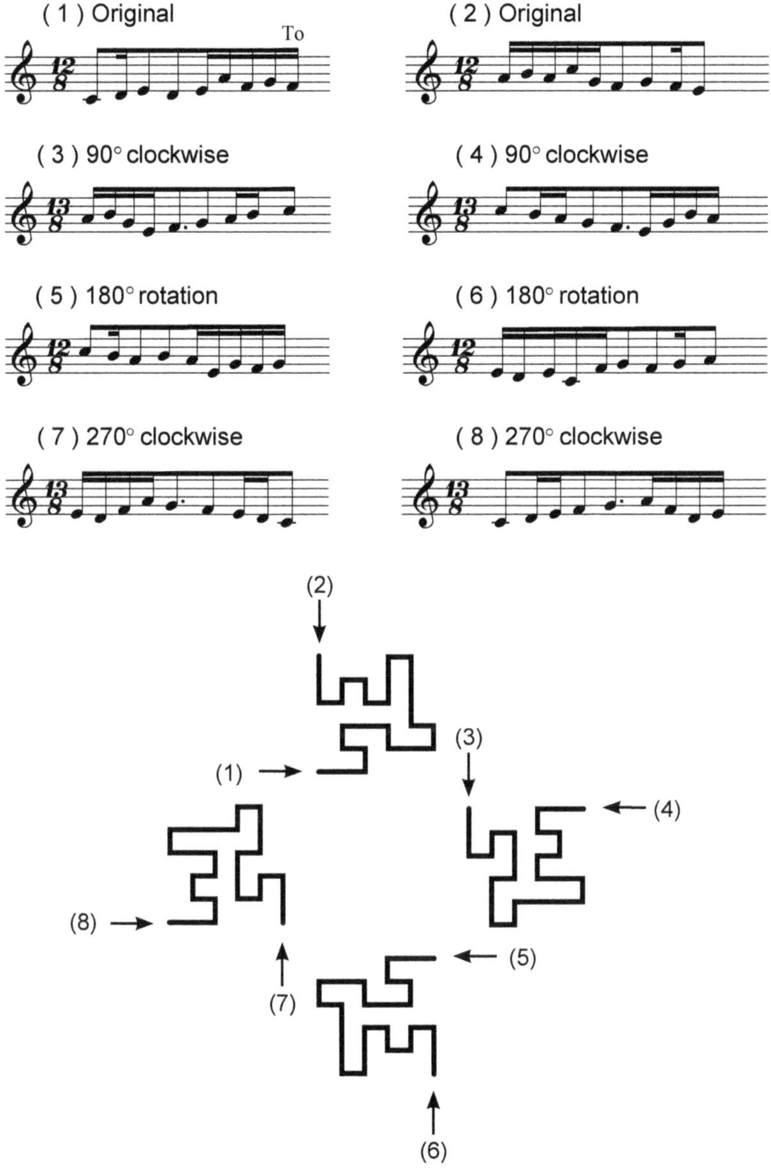

Tom Johnson. Piece No. 14 from the cycle *Rational Melodies* (1981)

Johnson's piece for piano No. 14 from his cycle *Rational Melodies* (1981) is an example of how the L-system is applied to musical composition. The structure of this piece is based on an endless number sequence (according to the formula *n* – >*n*, *n* + *1*, *n*), and is formed according to a repetitive cumulative counting, where each number in the new step is adequately exchanged with three new members.[164]

According to the composer, he created a number sequence where each 0 was changed to 010, and 1 was changed to 121, and 2 was changed to 232, and so on. Each number was matched with a respective sound. In this manner, an endless melody was composed (more see Johnson 2001). The analysis of this piece revealed that in the piano part the number sequence was transferred adequately by choosing the tone c^1 as the starting point for the combination 010. Respectively, the number combination 121 was intoned from the tone g^1, 232 – from c^2 and so on. Based on these observations, a graph of the beginning of the composition was created in which the L-system eternal sequence model was recreated (see Figures 101 & 102).

Figure 101. Arrangement of endless melody, when c – *0,* d – *1,* e – *2,* f-sharp – *3 etc.*

n – > *n*, *n* + *1*, *n*
0 (*c*)
010 (*c–d–c*)
010121010 (*c–d–c–d–e–d–c–d–c*)

010121010121232121010121010 (*c–d–c–d–e–d–c–d–c*...)

Figure 102. Johnson, piece No. 14 from the cycle Rational Melodies. *L-system application to the musical melody, when* cdc – *010,* gag – *121 and so on*

164 A similar occurrence of such musical composition was presented in the peculiarities of Johnson's piece for piano *Mersenne numbers*, pp. 133–136.

Gary Lee Nelson. *Summer Song* (1991)

The manner in which the L-system was applied in Johnson's piano composition is a rather direct application of this principle in music. The musicologist Martin Supper noted this, arguing that "the use of compositional rules derived from L-systems is usually not enough to generate a structurally coherent composition" (Supper 2001: 50). The American composer Gary Lee Nelson (born 1940) sought to avoid the "literal" transference of formal curves into musical space and modified them in a variety of ways. He changed the angle of rotation, stretched the curves, or pressed them together, and by so doing, changed their regular shape.

Nelson's pieces *Summer Song* for flute (1991) and *Goss* for violin (1993) reveal an original manner of using the L-system curve. In the first instance, the composer applied a modified Hilbert curve to form the musical structure. And in the second piece, he adjusted the graphics of a Gosper curve (Nelson 1996: 1). According to the composer, while composing *Summer Song*, he applied a few changes to the graphic prototype: firstly, he changed the regular angle of rotation (90°) to the angle of 101°. Then, the Hilbert curve's variant was laid out on a plane and transcribed into a musical scale with the intervals set out as M2 – M2 – m3 (see Figure 103). Here Nelson himself applied the limits of duration and pitch; he chose the sixteenth note as his smallest rhythmic unit and the half note as his longest. He chose the pitch c^1 as the lowest tone on his scale and *f-sharp*^{3} as the highest. It was noticed that the change of the curve's rotation opened up an opportunity to expand the range of rhythm up to 20 different rhythmic values (while the steps in the original Hilbert curve can only be expressed in three different units of duration). The recreation of a modified Hilbert curve in the first measures of the piece (see Figure 103) shows that the composer was not entirely accurate when he transposed the musical notes from the graph. Though the fine contours of the curve offered the sensitive microtonal relationships, Nelson applied the chromatic scale only.

Figure 103. Nelson, Summer Song. *Compositional process: 1) example of Hilbert curve; 2) modification of the curve changing the angle of rotation; 3) curve's variant laid out on a plane; 4) transcription of the curve into a musical scale; 5) recreation of Hilbert curve at the beginning of flute part*

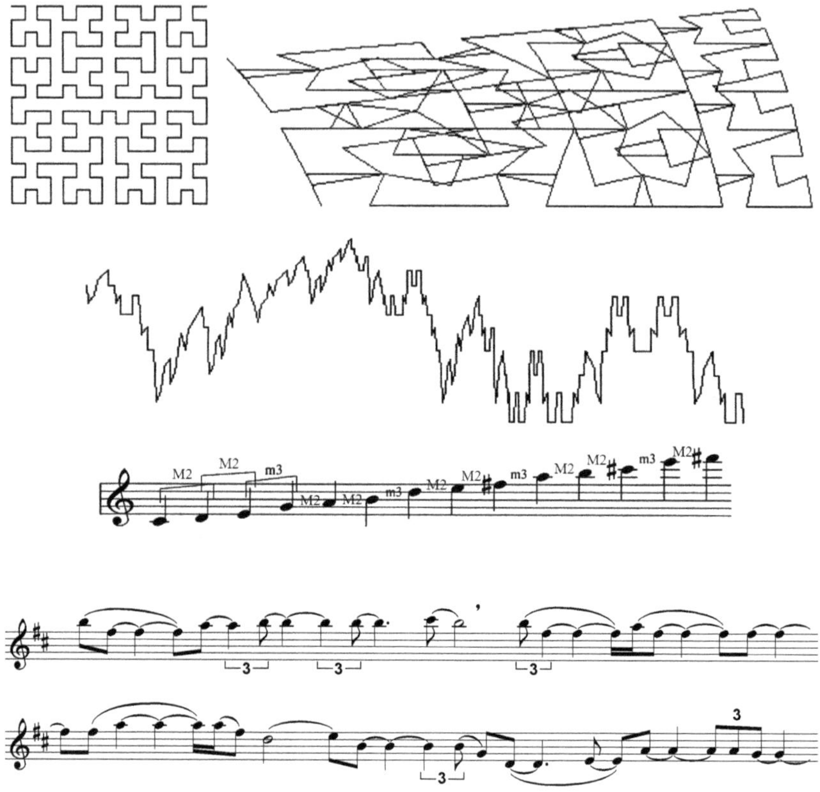

1.3. The Numericalization of the Musical Score

Morton Feldman. *IXION* (1958)

Morton Feldman was interested in the idea of indeterminacy and its expression in music. As one of its possibilities, Feldman chose the approximate pitches in a musical score. In Mello 1995, he referred to this as a liberation of sounds. On a practical level, the composer brought this idea to life in the compositions he created between 1950 and 1967. Here, writing on graph paper, the composer applied an indeterminate notation of pitch.[165] We find an analogical instance in Cage's score *Music for Carillon No. 1* (1952). Cage marked dots on graph paper that were approximate indications of pitch. However, later Feldman noted a failing in this method. Freedom was given not only to the choice of pitch, but also to the performer, while the composer did not agree with allowing performers to improvise.

Here, I would like to provide a detailed analysis of Feldman's *IXION* for chamber orchestra[166] (1958). Its graphic expression represents the application of mathematics on the creative process, because rather than writing the score using notes, the composer used numbers, which determined the number of notes. The *IXION* prototype can be seen in his earlier piece *Extensions 1* (1951). The symbolic notation in this piece is written out in numbers on graph paper as well. The score is covered with numbers, which indicate the number of notes, while lines on the graph paper denote the register. However, a choice of pitch was left to the discretion of the performer (Pritchett 2001: 690–6). At the same time, the use of number symbols as notes can be found in Feldman's 1951–3 "graphic" compositions *Intersections I, II*, and *III*.

At first glance the chaotic number codes scattered across the pages of *IXION* appear to look like improvisation. It would seem that Feldman wanted to eliminate any constructive moments from the process of musical composition, leaving only the phenomenon of chance. This would create a composition that did not have a defined pitch, a composition that every time suggested an opportunity for a new interpretation of the same numerical score. However, it is possible to see certain principles of the inner order of *IXION*. For example, the vertical alignment of the graphic score is made up of six rows of small square cells. The rows are divided down the middle by a bold line. According to the composer, the three lines set the approximate levels of register – the high point, the middle, and the low. The numbers written inside the squares determine how many times the tones should be played. Every 10 squares, the following set of Roman numerals, I–II–III–IV–V, is applied and repeated 14 times (the last time this process is incomplete):

I–II–III–IV–V I^{1}–II^{1}–III^{1}–IV^{1}–V^{1} I^{2}–II^{2}–III^{2}–IV^{2}–V^{2} … I^{14}–II^{14}–III^{14}–IV^{14}

The chaotic manner in which the numbers were written in the squares has an isomorphic basis, because the *IXION* two-part structure is characterized by a precise repetition in the middle. The contents of the section *A* (I^{9}–II^{10}) is accurately repeated in the section A_1, from II^{10} to IV^{11}. The following section, *B* (IV^{11}–V^{11}) is precisely repeated in the section B_1, from I^{12} to II^{12} (see Figure 104).

---

165 Feldman's compositions *Atlantis* (1959), *Out of 'Last Pieces'* (1961), and *In Search of an Orchestration* (1967) are written with indeterminate notation.

166 In 1960 this composition was adjusted to include two pianos.

The logical basis is inherent in the record of respective numbers as well. In total, the squares of the graph paper are filled in with 18 number symbols (from 1 to 18). I put those numbers into a table and then applied the data to draw a graph that would chart the keystroke of the tones: on the *x* axis the numbers 1 to 18 were marked; on the *y* axis the quantity of appearance of each of these numbers was noted. The obtained result revealed the consistent tendency to move from the largest frequency of the keystroke to the smallest. The figure 105 shows that the number one appears 315 times, while a set of 17 or 18 notes was played only once in the musical score.

Figure 104. Feldman, IXION. *Isomorphism in the middle section, from I^9 to II^{12}*

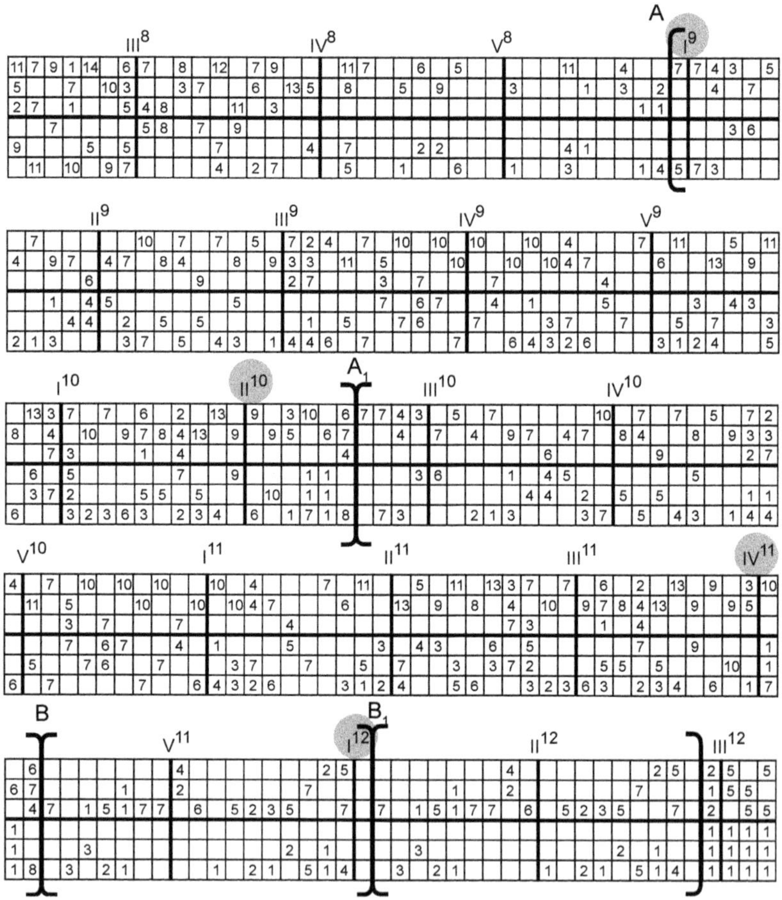

Figure 105. Feldman, IXION. *Table and diagram of appearance of numbers from 1 to 18, recorded on the graph paper*

Rows Numbers	I	II	III	IV	V	VI	Total
1	33	35	44	53	67	83	**315**
2	22	26	31	12	23	42	**156**
3	26	34	25	14	27	49	**175**
4	20	23	23	23	31	46	**166**
5	45	59	35	26	39	41	**245**
6	37	20	19	14	11	27	**128**
7	52	35	30	27	28	26	**198**
8	16	26	5	3	4	7	**61**
9	21	21	11	8	4	2	**67**
10	15	15	8	2	5	4	**49**
11	23	15	3	3	4	2	**50**
12	6	3	2				**11**
13	9	7	1				**17**
14	12	1					**13**
15	2						**2**
16	2	1					**3**
17	1						**1**
18	1						**1**
	334	321	237	185	243	329	**1658**

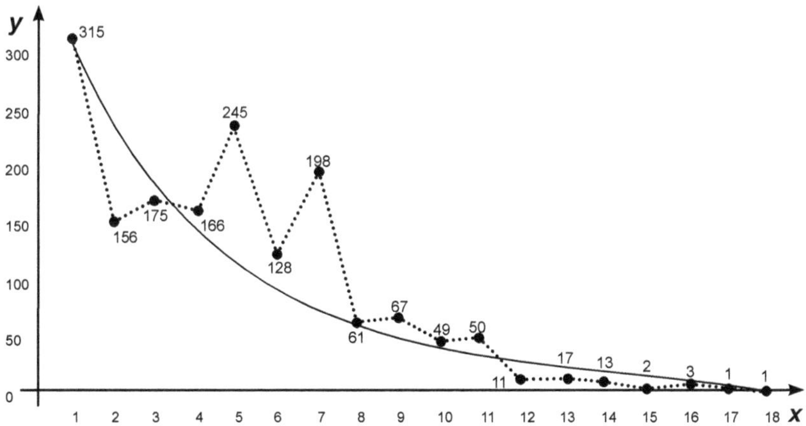

2. Implications of Modern Mathematical Theories

When studying how the elements of modern mathematics are applied to the peculiarities of contemporary music, one more often than not comes across the concept of the algorithm. It is no coincidence that the algorithm is associated with the practice of computer-generated musical composition. The application of generated models allows the composer to transcribe complex mathematical processes into musical tones. All of that is usually done by special computer programs written to handle this task. The analysis of music generated by algorithms, that is the recreation of certain procedures, also most of the time can only be done by computer. Additionally, many computer-generated/electronic music compositions are not written using traditional notation. They are expressed using formulas, symbols of the commands for programming. This phenomenon in the field of musicology presents the researcher with the challenges of an entirely different way of conducting an analysis of a musical composition.

2.1. The Practice of Algorithmic Music and Computer-Generated Composition

The concept of the algorithm[167] has been tied with the field of musical composition since the beginning of the 20th century. However, Gerhard Nierhaus notes examples of algorithms used in music as far back as Bach's *Die Kunst der Fuge*

167 The etymology of the concept of "algorithm" is connected with the Greek word *arithmos* (number) and the 9th century mathematician and astronomer Muhammad ibn Musa al-Khwārizmī (Khwarizmi), his surname's Latin form, Algoritmi. Al-Khwarizmi's treatise *Kitab al-Jabr val-Mukabala* (c. 820) presented the counting with Indian numerals and greatly contributed to the establishment of the so-called Arabic numeration in European mathematics; the concept 'al-Jabr' was exchanged with an 'algebra' (according to Cope 2000: 1; Supper 1997: 63).
This scholar's Indian method of counting slowly took root in Europe. Although, according to Gerhard Nierhaus, the number zero, 0, created problems for this system of counting to be accepted in the European world. In Antiquity there was no 0, because 0 was interpreted as nothing, chaos. Christian theology, which was influenced by Aristotle's philosophy, also negated the existence of nothingness (Nierhaus 2009: 14). Al-Khwarizmi's treatise in 1120 was translated into Latin as *Algoritmi de numero Indorum*. The author's name was written as 'Algorismus'. One more treatise in which

or Schönberg's dodecaphonic system (Nierhaus 2009: 1). According to Kristine Burns, the concept of the musical algorithm is close to the use of a medical algorithm to pinpoint a clinical diagnosis, because of its gradual process and the chain of interconnected questions and answers (Burns 1994: 2). This allows one to study the influence of algorithm on various styles and genres of professional music. Adam Alpern (Alpern 1995: 1) suggested descriptions of the process of algorithmic musical composition is as follows:

> The area of automated composition refers to the process of using some formal process to make music with *minimal human intervention.*[168]

This statement suggests the narrowing of this problem to the contemporary active use of the computer as a creative tool within the field of electronic music. Most of the time, computer-generated musical compositions are referred to as algorithmic. This is evidenced by terminology, such as "computer aided" or "computer assisted" composition (Cope 2000: 2). Electronic compositions are differentiated into computer composed and computer realized music as well (Dodge & Bahn 1986: 187).

The growing interest in the composing of computer-generated algorithmic music may be motivated by the fact that contemporary artists, with the aids of computers, have been able to reach a level that is equal to that of a highly mathematized tonal space. There are now a variety of opportunities to adapt, for example, nature's mathematical principles of evolution or algorithms of genetics to the field of musical composition. Complex computer programs are written to accommodate this interest. The first program to generate musical compositions was written in the 1950s by Lejaren Hiller and Leonard Isaacson. The first algorithmic computer-generated composition, *Illiac Suite* for string

---

the term *algorismus* can be found is the French scholar Alexander of Villedieu's *Carmen de Algorismo* (The Poem about Arithmetic, beginning of the 13th century). Later, the Greek version of *algorithmus* became the more preferred version. It meant the definition of controlled procedures. Today the concept of algorithm is defined as a set of finite rules, or a sequence of operations with the purpose of reaching a concrete goal.

The principle of the Roman numerals is typical of algorithms, because here the numbers are written using several symbols, which indicate a certain sequence of action. Aristotle's logical deduction theory may be described as an algorithmic procedure as well. Here Gerhard Nierhaus provides an example (Nierhaus 2009: 16): the main supposition is that man is mortal; the second supposition is that Socrates is a man; therefore, Socrates is mortal.

168 Italics are mine – *R. P.*

quartet (1955–6), was composed by these same authors. The interval sequence of this quartet used Markov chain algorithms. The rule-based principle of statistical processes was applied to program computer-generated music (Mason et al. 1988: 794; Cope 2008: xiv). Later examples are *Algorithms* (1968) by Hiller and Isaacson, Xenakis' experimental pieces from 1971, and his anthological instance of computer composition, *Gendy 3* (1991), where the tonal synthesis was generated by GENDYN (GENeration DYNamique, further described as Dynamic Stochastic Synthesis).

Often the musical concepts "algorithmic composition" (AC) and "computer assisted composition" (CAC) are used synonymously. Still, it is important to precisely differentiate the use of algorithms in music. The creative principles of Mozart's dice game (*Würfelspiel*) can be applied to algorithmic processes, as well as the Renaissance mensural system, the various types of isorhythmic motet and the canon, in addition to other examples of music where the results were achieved through the use of certain rule-based operations.

The composition of computer-generated music based on algorithms became more intense from the second half of the 20th century onwards, encouraging one not only to come to certain analytical generalizations, but also influencing more complex directions in musical analysis. This is because the concept of the algorithm takes up an especially wide spectrum of creative procedures. On the one hand, when analyzing the concept of algorithmic composition, it is important to differentiate between imitative work (for example, imitations of a specific style) and an original musical composition generated by an algorithm.

On the other hand, according to Miller Puckette (2006: ix), a purely technical differentiation is applied. That is "computer generated music" (CGM, Denis Baggi's concept) and the aforementioned "computer assisted composition" (CAC):

- computer generated music, CGM, is equated with a synthesized, processed and/or designed sound. The computer takes on the role of the music instrument;
- computer assisted composition, CAC, is entrusted with the task of performing complex mathematical calculations, using complex algorithms, that is, functions that formerly were attributed to human thinking or creative acts.

We may expand Puckette's description by David Cope's triad (2008: ix–x):

- first, "computer generated sound" (CGS), that is close to the sound generated by a synthesizer;

- second, "computer generated assistance" (CGA). The computer is an aid that is used to organize the various elements of music and the parameters of musical language;
- third, "computer generated composition" (CGC) that is close to the aforementioned CGM.

The variety of algorithmic processes in music can be grouped according to characteristic technological principles as well. The result of their application is a unique reproduction of the original musical material (see Cope 2008). The most popular are:

- sonification (sound processing, the use of non-sound data), meaning a computerized translation into sound;
- a variety of mathematical operations, the use of calculations, the choice of the mathematical operation for a musical prototype, or the composition of new musical material generated by applied musical data analysis (*Data-Driven*); for example the analysis of already existing musical compositions.

The adaptation of algorithms to the process of musical composition can be differentiated according to whether the algorithm is only one of the tools of composition, or whether an algorithm is systematically applied to the musical language, structure of the composition (Nierhaus 2009: 261). In the first instance, the given algorithm is controlled by the composer and "taken care of." The algorithm influences the different musical parameters. An example would be pitch, rhythm, dynamics, and their combinations. The second instance would be where the algorithm, or a combination of algorithms, is given a more independent function, one that would encompass the entirety of compositional codes. This provokes a debate within the realm of creative dimension – does the algorithm take over the role of the composer completely, or at least in part. David Cope has been thinking along these lines. He sees two aspects of the composer's intention in the creative production of computer-generated music. One group, for example, Bryan Ferneyhough, composes music in a traditional manner, but uses the computer as an additional tool for composition. This group of composers experiments with certain models, but modulates the final result according to their own "taste" (meaning, giving the work a human touch with the space of the computer-generated composition). The second group, like Gottfried Michael Koenig, in most of his compositions uses the computer program as a complete method for problem solving, for organizing all of a work's parameters, and to control the musical material (Cope 2008: x–xii).

While trying to generalize how algorithmic actions are targeted to create music, in the practice of music, I would divide their ideas into two directions, as follows:

- The universal application of the concept of "algorithm." This encompasses a variety of determined (set, defined) processes, a variety of constructive actions, which are performed by the composer himself, while holding to respective rules.

Part of these algorithms function mechanically. They are easy to distinguish in tonal material. All the examples of aspects of musical composition based on certain mathematical elements, from all of the epochs we have discussed thus far – numbers, proportions, progressions, and so on – could also be attributed to this direction of algorithmic procedures. We can also see the algorithmic nature in Pythagoras' theory of tone-number equivalents. In the opinion of Järveläinen (Järveläinen 2000: 1), algorithmic aspects have a unique influence on formal compositions from the Middle Ages that were created according to the graphs of Guido of Arezzo, on rhythmic models of 15th century isorhythmic motets, or in Dufay's compositions, which show the Golden Ratio in their inner structure. The performance marks for Renaissance mensural canons are also interpreted by applying algorithms. In this case, the composer is merely the author of the initial motif, or the core of the composition, from which the composition is constructed, and of a complex of rules. Schönberg's twelve-tone series, which Webern expanded into total serialization, according to John A. Maurer, is noted for controlling absolutely all musical parameters, maximally abstracting the composition process (Maurer 1999: 2). For this reason the matrixes (series) that are made up of the dodecaphonic principle encourage to name them as algorithms.

- Complex mathematical algorithms, as well as algorithms from other fields of science, are applied to the composition of contemporary music. However, their development and application is difficult to notice, and they are often only brought to life in the computerized realm.

The composer concentrates on the complicated process of composing music, which is based on complex mathematical procedures. An example would be the experimental music composer Charles Dodge's composition *The Earth's Magnetic Field* (1970). The number sequences of the movement of the Earth's gravitational field was used as the prototype for the tonal fabric of this piece. The computer converted the numbers into tones (Alpern 1995: 1). In Dodge's composition *Profile* (1984) the pitch, rhythm, and amplitude were influenced by an adaptation of the 1/f noise algorithm. According to Järveläinen (Järveläinen 2000: 10), the principles of Markov chains are especially appropriate for generating musical melodies. Another field that inspires contemporary composers are genetic algorithms. The reason for this choice is that the structures of genetic DNA (Deoxyribonucleic acid) and musical structures are similar to linear sequences of elements that create complex combinations. Additionally, DNA opens up to noise signals that are similar to 1/f. John Dunn used the characteristics of amino acids (one of the DNA components) – molecule masses and the dimensions of clay material – as the basis for his musical tones as well as for other musical parameters. Xenakis experimented with the implications of stochastic processes on music. Nelson chose the chaos and fractal theories as the prototype for his computer-generated compositions *Fractal Mountains* (1988) and *The Voyage of the Golah Yota* (1993). Hanspeter Kyburz (born 1960) used recursive sequence algorithms for his composition *Cells* for saxophone and ensemble (1994).

2.2. Tonal Adaptations of Complicated Mathematical Processes

The practice of computer programmed algorithms in contemporary music generalized from a few different sources (see, for example Burns 1994, Dodge 1988, and Järveläinen 2000) can be examined according to a choice of respective mathematical phenomena. In order to compose music using the computer one may apply the following:

- stochastic processes (chance, probability, Markov chains);
- chaotic processes that encompass fractal structures;
- structural linguistics analysis (grammatical processes);
- rule-based, causal action processes;
- artificial intelligence, AI, processes that are based on the right choice for the next step.

Nierhaus makes more minute differentiations in complex mathematics in music. He denotes eight algorithmic principles for musical composition and/or analysis, thus elaborating the above mentioned list with a few more aspects (more see Nierhaus 2009). That is:

- Markov chains/model;[169]

169 The model is based on the probability principle, when the likelihood of a future act is based on one or a few acts that have already taken place, which influence the later process, which is called the order. In the beginning part of this process there is a high degree of uncertainty (a lack of stability). In the later stages of the chain the certainty becomes stronger (more predictable).
The Markov chain theory was developed by the Russian mathematician Andrei Markov (1856–1922). He tried to define the special characteristics of literary lan-

- the principles of generative grammars;[170]
- analogies of transition networks;
- the adaptation of the chaos, self-similarity principles (Verhulst equation, various theories of noise,[171] chaos systems, Cantor set, Mandelbrot set, the Koch snowflake and other fractals);
- genetic algorithms;
- cellular automata;[172]
- artificial neural networks;[173]
- artificial intelligence.[174]

The origins of **generative grammar in music**[175] can be identified already in the early 20th century with Heinrich Schenker's (1868–1935) *Ursatz* theory. Later the

guage. While analyzing Pushkin's *Eugene Onegin*, he determined when and where in Pushkin's text the same vowel is repeated, where and when one vowel is followed by another, where and when a consonant comes after a vowel (Ames 1989: 175–6).

170 In the 1980s, after Noam Chomsky created the model theory, which is based on linguistic and hierarchical principles, the study of generative grammar became popular in musicology, and its principles were applied to the analysis of musical structures.

171 For example, researchers Richard F. Voss and John Clarke, after studying a variety of noise forms (white, pink, etc.), established that analyzed examples of music from various epochs and styles often have frequencies that match the $1/f$ noise spectrum (more see Voss & Clarke 1978).

172 Cellular automata algorithm is often compared to the genetic algorithm theory, although the only thing they have in common is their terminology, and not their principles of functioning (Nierhaus 2009: 201).
The principle of cellular automaton is closer to the L-system. In order to generate music, most of the time main cellular automaton rules are not applied, but rather a method of graphic representation, a musical cartography, where the cells move according to a certain order in a net (Nierhaus 2009: 202).

173 Algorithms of artificial neural networks are usually used to compose music, and not to analyze music. According to Nierhaus, it is only possible to match a few artificial network types, because these algorithms typically have a weak processing ability (Nierhaus 2009: 221).

174 As Nierhaus notes, the principles of artificial intelligence, fundamentally, are best suited to generate algorithmic music in the Baroque style (in order to create Baroque melodies, four-voice polyphony, and so on). This algorithm can also be used with other algorithms, like Markov chains, the principles of generative grammar (Nierhaus 2009: 200).

175 This type of analysis lends itself especially well to the study of ethnic music. In this manner, gamelan, Swedish folk music, the music of Inuit (Canada) and Yupik (Siberia, Alaska), Indian musical tablas and other structures were analyzed. The principles of generative grammar were used to attempt the classification of chord sequences in jazz music, the logic of improvisation (Nierhaus 2009: 100–7).

generative grammar model was successfully adapted in Fred Lerdahl's and Ray Jackendoff's research. In order to reduce the composition to its most essential tones, they strove to purify the hierarchy of layers of musical structure by analyzing the relationship of groups (when a piece is divided into motifs, phrases, and sections), metric structure, and the reduction of a duration, which is connected with pitch, group and metric indicators. Lelio Camilleri's research was connected with the principles of generative grammar, which he applied to the analysis of few songs from Schubert's vocal cycles Op. 23, 25, and 89. Using the derived data of initial phrases, Camilleri generated this style's melody copies (Nierhaus 2009: 94, 99). Together with a group of scientists, Mario Baroni created a few computer projects that analyzed examples of classical music or composed music in the chosen style according to the principles of generative grammar. The program Melos 2 was adapted to analyze the structure of Lutheran chorales; the program Harmony generated a bass and a harmonic vertical for a given melody (more see Baroni et al. 1982).

In the visualization of the generative grammar theory it is possible to see analogies with elements of chaos theory. For example, the graph of a prolonged tree for Bach's chorale *Christus, der ist mein Leben*, BWV 95, expresses the relationship between the chorale's tones, the tendencies of tone groupings, and hierarchy. Without a doubt, the image resonates with one of the chaos theory graphs – that of the Verhulst model that shows a diagram of population growth (see Figure 106).

Figure 106. Bach, chorale Christus, der ist mein Leben, *BWV 95. Music score visualization as a prolonged tree (reproduced from Lerdahl 2001: 22)*

Genetic algorithm principles are essentially similar to the traditional process of music composing. According to the algorithm, the initial model experiences its structural modifications, mutations, conflict, and other actions in order for biological genes to form; the adequacy of the result is compared with other generative models, and so on. In the sound space, it resembles the development and modification of the musical material, or the principles of inversion, retrograde, and other transformations.

Andrew Horner and David E. Goldberg were among the first to write about thematic bridging in the generating of music. They give an example how, according to the principles of genetic algorithms, out of the chosen five-tone motif *g-flat – b-flat – f – a-flat – d-flat* it is possible to generate a certain melody (Horner & Goldberg 1991: 479–80):

- firstly, the last tone is discarded, and the result is the four-tone motif:
  g-flat – b-flat – f – a-flat
- second, the new motif's tones change places, for example:
  b-flat – f – a-flat – g-flat
- third, from the last motif the final tone is discarded:
  b-flat – f – a-flat
- fourth, the first tone of the resulting motif is changed, for example:
  e-flat – f – a-flat
- fifthly, the tones once again change places, for example:
  f – a-flat – e-flat
- sixthly, the result is achieved – a sequence of all the tones:
  g-flat – b-flat – f – a-flat – b-flat – f – a-flat – g-flat – b-flat – f – a-flat – e-flat – f – a-flat – f – a-flat – e-flat

The principles of genetic algorithms can be applied to the analysis of classical music as well. For example, Michael Towsey et al. used the computer to analyze Renaissance compositions and popular music, as well as children's songs. They analyzed a total of 36 pieces and put together a chart to compare all of them according to 21 features – pitch, tonality, melodic relief, rhythm, repeating models, motifs, as well as a complex of other qualities (see Towsey et al. 2001).

The use of the **Markov model and the practice of stochastic music** to compose music can be explained as the aims of composers in the creative process to smooth any cognitive possibility. Xenakis, protesting against the strict serialized control of tone, chose the more common concept of "stochastic music" as a scholarly synonym of chance and applied the distribution of probabilities to musical composition. For example, in the composition *Pithoprakta* (1955) probability processes determined the duration of tones that were formed based on the kinetic theory of gas, while the tonal changes revealed the analogies of the stochastic process (Järveläinen 2000: 3).

The application of the probability process can be seen in Cage's declaration seeking to free himself from an individual approach that influences his creative process (Pritchett 2001: 690–6). Already at the beginning of the 1950s, when Cage read the *I Ching* (*Book of Changes*), those texts inspired him to take an interest in probability theory. The composer used the process of tossing coins in his creative process. An intermediary, an oracle, chose the pitch from the charts that he created. The rather coincidental sound in his piece *Reunion* (1968) is based on a chess game; the tones are created by a photoreceptors attached to the chessboard – when the player lifts the chess piece, a tone is heard.

This method of composing music based on assembling chance elements is specific for the rhythmic picture of Cage's *Ryoanji*. The chart of rhythmic sequences in the percussion part and corresponding number sequences display that no one rhythmic motif is repeated, because at least one element constantly is changed and moved to another place (see Figure 107).

Figure 107. Cage, Ryoanji. *Table of the rhythmic plan*

Meter 12 ♩ (16 m.)	Meter 13 ♩ (16 m.)
♩𝄽𝄽♩𝄽𝄽♩♩𝄽𝄽♩𝄽 1–2–1–2–2–2–1–1	♩𝄽𝄽♩𝄽𝄽♩𝄽♩𝄽♩𝄽𝄽 1–2–1–2–1–1–1–1–1–2
♩𝄽𝄽♩𝄽𝄽♩♩𝄽♩𝄽𝄽 1–2–1–2–2–1–1–2	♩𝄽𝄽𝄽♩𝄽♩♩𝄽𝄽♩𝄽𝄽 1–3–1–1–2–2–1–2
♩𝄽𝄽𝄽♩𝄽♩♩𝄽𝄽𝄽♩ 1–3–1–1–2–3–1	♩𝄽♩♩𝄽𝄽𝄽♩𝄽𝄽♩𝄽𝄽 1–1–2–3–1–2–1–2
♩♩𝄽♩𝄽♩𝄽𝄽♩𝄽𝄽𝄽 2–1–1–1–1–2–1–3	♩𝄽♩𝄽♩𝄽𝄽𝄽♩𝄽𝄽𝄽♩ 1–1–1–1–1–3–1–3–1
♩𝄽𝄽𝄽♩♩𝄽♩♩𝄽𝄽𝄽 1–3–2–1–2–3	♩𝄽𝄽♩𝄽♩♩𝄽𝄽♩𝄽𝄽𝄽 1–2–1–1–2–2–1–3
♩𝄽𝄽𝄽♩𝄽𝄽♩𝄽♩𝄽♩ 1–3–1–2–1–1–1–1–1	♩𝄽𝄽𝄽♩𝄽♩♩𝄽♩𝄽𝄽𝄽 1–3–1–1–2–1–1–3
♩♩𝄽𝄽𝄽♩♩𝄽♩𝄽𝄽𝄽 2–3–2–1–1–3	♩𝄽𝄽𝄽♩𝄽♩𝄽♩𝄽♩𝄽𝄽 1–3–1–1–1–1–1–1–1–2
♩𝄽𝄽𝄽♩𝄽♩𝄽𝄽♩𝄽♩ 1–3–1–1–1–2–1–1–1	♩𝄽♩𝄽♩𝄽𝄽♩𝄽𝄽𝄽𝄽♩ 1–1–1–1–1–2–1–4–1
♩♩𝄽𝄽♩♩𝄽𝄽𝄽♩𝄽𝄽 2–2–2–3–1–2	♩𝄽𝄽♩𝄽♩𝄽𝄽𝄽♩♩𝄽𝄽 1–2–1–1–1–3–2–2
♩♩𝄽𝄽𝄽♩𝄽𝄽𝄽♩♩𝄽 2–3–1–3–2–1	♩𝄽𝄽𝄽♩♩𝄽𝄽♩♩𝄽𝄽𝄽 1–3–2–2–2–3
♩♩𝄽♩♩𝄽𝄽♩𝄽𝄽♩𝄽𝄽 2–1–2–2–1–2–1–2	♩𝄽𝄽𝄽♩𝄽♩♩𝄽𝄽♩𝄽𝄽 1–3–1–1–2–2–1–2
♩♩𝄽𝄽♩𝄽♩𝄽𝄽𝄽♩𝄽 2–2–1–1–1–3–1–1	♩𝄽♩𝄽𝄽𝄽♩𝄽♩𝄽𝄽𝄽♩ 1–1–1–3–1–1–1–3–1
♩𝄽𝄽♩𝄽𝄽♩𝄽𝄽♩𝄽♩ 1–2–1–2–1–2–1–1–1	♩𝄽𝄽𝄽♩𝄽𝄽♩𝄽𝄽♩𝄽♩ 1–3–1–2–1–2–1–1–1
♩♩𝄽♩♩𝄽𝄽𝄽𝄽𝄽♩𝄽 2–1–2–5–1–1	♩𝄽𝄽𝄽♩𝄽♩𝄽𝄽♩𝄽♩𝄽 1–3–1–1–1–2–1–1–1–1
♩𝄽♩𝄽𝄽𝄽♩𝄽♩𝄽𝄽𝄽 1–1–1–3–1–1–1–3	♩𝄽𝄽𝄽♩𝄽♩𝄽♩𝄽♩𝄽𝄽 1–3–1–1–1–1–1–1–1–2
♩𝄽♩𝄽♩𝄽𝄽♩𝄽𝄽♩𝄽 1–1–1–1–1–2–1–2–1–1	♩𝄽𝄽♩𝄽𝄽♩𝄽𝄽♩𝄽𝄽♩ 1–2–1–2–1–2–1–2–1

Meter 14 ♩ (21 m.)	Meter 15 ♩ (24 m.)
1–1–1–3–1–2–2–3	1–3–1–3–1–2–2–2
2–2–1–2–1–3–1–2	1–3–1–3–1–2–1–1–1–1
1–1–1–1–1–3–1–3–1–1	2–2–1–2–1–3–1–3
1–1–1–3–1–3–2–2	1–3–1–2–1–2–1–3–1
1–2–1–2–1–3–2–2	1–2–1–1–1–3–1–3–1–1
1–4–1–1–1–2–2–2	1–5–1–3–2–1–1–1
1–3–1–3–1–2–2–1	1–2–1–3–1–1–1–2–1–2
1–1–1–2–1–2–1–1–1–3	1–3–1–3–1–2–1–2–1
1–2–2–3–1–1–1–3	1–2–1–1–1–3–1–2–1–2
1–1–1–2–1–3–1–2–1–1	1–1–2–4–1–2–1–3
2–1–1–5–2–3	1–1–1–3–1–3–1–3–1
1–3–1–2–1–1–1–2–1–1	1–3–1–3–1–1–1–2–1–1
1–3–1–2–1–1–2–3	1–3–1–2–1–2–1–1–1–2
1–1–2–1–1–3–1–4	2–5–1–3–1–2–1
1–3–1–3–1–1–1–2–1	1–3–1–3–2–1–1–3
1–1–1–2–1–2–1–2–1–2	1–2–1–2–1–3–2–3
1–4–1–1–1–1–1–2–1–1	1–3–1–1–1–1–1–2–1–3
1–1–1–3–1–3–2–2	1–2–1–4–1–3–1–1–1
1–3–1–2–1–3–1–1–1	1–1–1–5–2–2–1–2
1–2–1–3–2–3–1–1	1–1–2–3–2–6
1–2–1–2–1–2–1–2–1–1	1–1–1–2–2–1–1–6
	2–3–1–4–1–1–1–2
	1–3–1–1–1–3–1–1–1–2
	1–4–1–3–1–2–1–1–1

The Markov chain theory is especially adequate for expressing stochastic musical processes. The application of such principles has become one of innovative means of composing music and analyzing it. The first time this method was used in music was in 1950, when Harry F. Olson applied it to analyze the songs of Stephen Foster. Wei Chai and Barry Vercoe used the principle of the hidden Markov model to analyze and compare pitch, interval structures, and duration data of the melodies of folk songs from Ireland, Germany, and Austria.

Based on Xenakis' experiments, in a musicological field, an attempt was made to differentiate several compositional methods of stochastic music (Theory of Contemporary Music 2005: 515). Therefore we can distinguish:

- free-form stochastic music (based on the probability theory);
- Markov's stochastic music (musical composition according to the Markov model);
- a musical strategy (the group theory is used in the composition).

However, according to Nierhaus, Markov chains could be applied to a one-dimensional symbol sequence analysis. Therefore, it is never completely adequate to thoroughly analyze a complex musical composition to evaluate its horizontal and vertical structure (Nierhaus 2009: 81).

The phenomenon of **chaos theory**. The concept of "chaos" is in opposition to the "cosmos" concept that represents harmony. However, according to Hiller, chaos is an organized system based on inside rules and brings an order to the creative music process. That is because "from the variety of endless possibilities, that is chaos, the orderly respective musical elements are selected" (Theory of Contemporary Music 2005: 514). The creation of chaos in contemporary musical composition can be mathematically based (using its mathematical formulas as compositional algorithms), and on abstract semantics. The latter description is suitable to rationally characterize Ligeti's creative work; for example, a logical construction of polymeter evokes the chaotic sound of the etude *Désordre* (additionally, the manifestation of quasi-fractal principle, a fractal imagery, will be presented in the analysis of Šarūnas Nakas' *Ziqquratu*).

Examples of the application of the complicated mathematical chaos principle in contemporary music is quite varied. It could be an expression of Granular Synthesis[176] which was first used by the Canadian composer Barry Truax (born 1947) in his piece *Riverun* (1986). Xenakis applied this phenomenon to his work *Gendy 3* (1991). The model for counting population expansion $f(X) = P \times X \times (1 - X)$, the so-called Verhulst equation,[177] has been applied to the practice of contemporary music as well. Nelson's composition *The Voyage of the Golah Iota* (1993) made use of this mathematical formula. The composer chose the numbers from 1 to 4 for the measurement P, for X – from 0 to 1, and using the computer, graphically recreated the equation data. The graphic visualization was transformed into a musical space. Additionally, to program this piece Nelson applied granular synthesis as well as the principles of genetic algorithms,

176 Granular synthesis is a method in computer-generated music sound that operates on microtonal structures; the techniques of analogs and selections are applied.

177 Mathematician Pierre François Verhulst (1804–1849) developed a chaos theory formula, the Verhulst equation; it is used to explain population growth through a genetic algorithm. The critical value is $P = 4$.

chaos actions that determined the repetition of elementary motifs and complex sequences (Nelson 1994: 1).

Research on contemporary music composition reveals the especially complex implication of **mathematical formulas**. For example, from Study No. 21 on Nancarrow began to apply complicated mathematical proportions to create the polyphonic tempo. The twelve-voice score of Study No. 37 is based on the elaborate mathematical relationships of the chromatic scale; in Study No. 33 the ratio of the movement of two voices was established by the formula written into the title, *Canon* $\frac{\sqrt{2}}{2}$, bringing together two different forms of the same number – the irrational numerator (square root of 2) and the rational denominator (the natural number 2). Study No. 40, which Kyle Gann called the transcendental cannon (Gann 1995: 200), has a title that was written down with two irrational numbers: *Canon* $\frac{e}{\pi}$. This could be described as the composer's intention to bring into conflict two opposite mathematical expressions, which were of a different nature, a dynamic and a static.[178]

According to American minimalist Johnson, his five-part piece for orchestra *Dragons in A* (1979) was composed based on the principles of the Dragon formula,[179] while the melody of the four-part piece for piano *Cosinus* (1994) is described as a mathematical structure played in a vertical from one to four voices (Johnson *Editions 75*). The formula for a Galileo number (a magnitude known in fluid dynamics) influenced the structure of Johnson's work *Galileo* (2000). In this composition five metal pendulums resemble Nancarrow's idea of a polymetric canon. The pendulums were hung in varying heights. As they moved, they created a sound of an increasingly more complex rhythmic counterpoint. The slowest pendulum was hung at the height of c. 4 meters, while others were hung according to the relationships defined by Galileo, 1/2, 2/3, 3/4 and 4/5. These relationships match the metronome marks of 20 – 25 – 26 $^{2}/_{3}$ –30 – 40.

The implication of complicated mathematical calculations are inseparable from Xenakis' musical exploration. An entire chain of mathematical functions are written out in the sketch for his composition *Achorripsis* for 21 instruments (1957). The composer posits that he relied on the probability theory, specifically the Poisson distribution with the formula $P_k = (\lambda^k/k!) \times e^{-\lambda}$ (where λ happens to be exactly the mean value of the Poisson distribution). When Xenakis transferred

178 The expression $e : \pi$ (where e is a natural logarithm basis) is still written using the numerical formula 2.7182818284… : 3.1415926536…

179 The Dragon curve (Dragon formula) is a special example of a fractal, self-similar curve, which is obtained from the so-called IFS, iterated function system.

Figure 108. Bach, canon cancrizans *from* Musikalisches Opfer, *BWV 1079. Example of musical palindrome and the Möbius strip (left)*

this sketch over into musical notation, the Poisson theory principle generated 196 different cells, which influenced the combinations of tempo and duration. In the composition seven timbre groups were used as well as 28 rhythmic units (196 : 7 = 28), which were laid out in a two-dimensional figure (Xenakis 1991: 29–31).[180]

The polyphonically constructed canon structure in certain instances can also be mathematically proven. For example, Bach's *Musikalisches Opfer* (BWV 1079) creates a refined musical palindrome example, the so-called crab canon (Latin *canon cancrizans*), after the *Thema Regium*. The design of this musical canon is analogical to the Möbius strip, which was named to honor August Ferdinand Möbius (1790–1868). Bach's musical theme embodies the phenomenon of infinity, a quality of the one-sided (mathematically – non-oriented) surface, because the melody plays harmoniously, performing both from its beginning and from the end, or while performing a simultaneous canon of its original and retrograde (that is, the beginning of the melody is the same as the end, and vice versa; see Figure 108). An example of a rather different canon perpetual is illustrated by a 11-measure period from Haydn's Menuetto of String Quartet in D minor *Fifths*, Op. 76 No. 2/Hob III: 76, where the musical material is played out in such a manner as though the end plays like a return to the beginning.

Another intriguing aspect of measurement, and an innovative path to musical composition that was offered to contemporary composers, is the **Vuza tiling rhythmic canons.**[181] This phenomenon is associated with the art of mosaics, which was especially perfected in the Byzantine and Islamic cultures and which master-

180 Other compositions by Xenakis that are based on the logic of the Markov chain are *Analogique A* for string orchestra (1959), *Analogique B* for sinusoidal sounds (1958–9) and *Syrmos* for eighteen strings (1959). In 1962 he completed *Morsima-Amorsima* for four instruments.

181 This method of composition is based on the Romanian mathematician Dan Tudor Vuza's (born 1955) canon theory.

fully adapts the rules of geometry, so that a certain area would be carefully filled in with the details of varying forms, which links everything into an impressive sight. In music, the same fragment of the melody, the motif, is laid out in several voices in different versions of duration (augmentation, diminution, etc.), so that the entire line would be filled without pauses (holes) and clashes in the vertical (doublings). The Vuza musical canons present composers with the complicated task of choosing the appropriate melodic part and to correctly set up its variants. Therefore, filling up the complex multi-voiced texture by hand is impossible most of the time and must be accomplished using computer calculations.

According to Moreno Andreatta, French composer Fabien Lévy's (born 1968) piece *Coïncidences* for orchestra (1999) is regarded to be the first musical composition that was composed according to the rules of the Vuza canon (Andreatta 2011: 53). As Lévy states, the adaptation of this mathematical phenomenon in music reveals the interaction between the polyphony and monody; but from a perceptual position it is not a traditional canon (Lévy 2011: 27–30).

The technology of Vuza canons is especially convenient for organizing a rhythmic picture. We see this in Tom Johnson's musical experiments: from 2003 onwards he wrote many compositions with the common title of *Tilework*. For example, in the composition *Tilework for Log Drums* (2005) a problem was "solved" as to how the variants of three-tone motif fill a six-voice score so that the same motif would play in a relationship of five different tempos, 5 : 4 : 3 : 2 : 1. A possible solution of this problem is illustrated in four different tempo graphs, 18 x 6 (see Figure 109).

Figure 109. Left: filling in the large square with the smaller squares of nine different sizes (two of the smaller squares, No. 6 and 8, are used twice); right: diagram for Johnson's six-voice composition Tilework for Log Drums *(the figure reproduced from Johnson 2011: 19)*

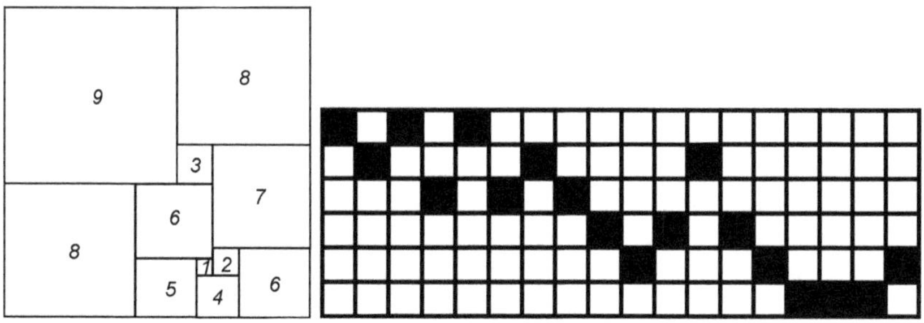

The mathematical direction of contemporary music is reflected in the use of the mathematical **group theories**. The author of transformational music theory, David Lewin, based his work on group theory when he researched the relationships of musical intervals as an expression of the transformational net. The group theory model is applied to the analysis of musical scales as well, because the transcription of the 12-tone system into a number sequence matches the module 12 principle; group C_{12} (a cyclic group of order 12) is made up of 12 of its elements, the tones (*c* – 0, *c-sharp* – 1 and so on).

Iannis Xenakis. *Nomos alpha* (1965)

The application of the group theory in Xenakis' composition for cello *Nomos alpha* (1965)[182] is first of all confirmed by the composer in his notes, referring to musician, philosopher and mathematician Aristoxenus of Tarentum (Ἀριστόξενος, born c. 375 B.C.), mathematician and author of group theory Évariste Galois (1811–1832) and his follower Felix Klein.[183] Thomas DeLio refers to this musical composition as a piece about time. In this piece, a unique dialectical time conception is combined with two elements – the discrete and perpetual (see DeLio 1980: 63–95). Evan Jones noted that in this composition Xenakis applied tonal effects to achieve extremes – "eeriness, aggression, and hyperactivity" (Jones 2002: 73). Robert W. Peck argues that *Nomos alpha* is not a stochastic composition *per se*, because it avoids stochastic moments, such as asymmetry, non-periodicity, and repetition (Peck 2003: 113).

While thoroughly describing this composition Xenakis presents a plan that details the creative process for *Nomos alpha*. The plan shows how he transformed a variety of mathematical structures into music and how he manipulated mathematical elements to match musical tones to cube rotations. According to the sketch, eight sound models (articulations and types of musical sound) were matched with eight vertices of the cube. The original model's number sequence 1–2–3–4–5–6–7–8 is changed according to the logic of group theory, turning

182 Later *Nomos alpha* was transcribed into *Nomos gama* (1968) for 98 orchestral musicians that were "scattered" throughout the concert hall. Such a positioning of the orchestra, according to the composer, was influenced by the development of *Nomos alpha*. Xenakis described the orchestral *Nomos gama* score as a generalization of *Nomos alpha* (Xenakis 1992: 236).

183 The annotation was published in Iannis Xenakis: *Nomos alpha*, Boosey and Hawkes, 1964.

around the cube in a different way. In this manner we get a different number sequence:

2–1–4–3–6–5–8–7, 3–4–1–2–7–8–5–6, 4–3–2–1–8–7–6–5, 2–3–1–4–6–7–5–8, 3–1–2–4–7–5–6–8 etc.

Then the cube is transformed into a tetrahedron (4 vertices) and its rotations are performed.

Moreno Andreatta argues that the tonal model is arranged according to Fibonacci rules, because the result is based on the relationship between the two members that come before. The computer additionally generates the indicators of dynamics and duration (more see Andreatta 2004 & 2007). However, when the composition is written down using traditional notation, and in the score there is at least minimal marks of the variations of cube rotations, then the principles of group theory remain hidden. It can only be reconstructed if one knows the pre-compositional scheme for *Nomos alpha*, i.e. the variations of cube rotations, as well as the compositional square, which were presented in Xenakis' study (Xenakis 1992: 219–36).

2.3. Fractal Theory Analogies in Musical Compositions

After Benoit Mandelbrot's[184] discoveries in the field of fractal geometry, this theory found its place in the imaginations of contemporary composers as a tonal realization of fractal geometry. The study of the expression of geometric fractals in music inspired new analytical works. According to two Swiss brothers, geologist Kenneth Hsü and musicologist Andrew Hsü, fractals are typical

184 Polish born mathematician Benoit Mandelbrot (1924–2010, his parents lived in Vilnius) tried to prove that according to the main theory of fractal geometry, fractals can be found in various phenomena in nature and in the structures of objects. Because of the findings of this scholar, it became possible to use mathematical relationships to express non-mathematical or non-geometric forms, such as clouds, mountains, trees, and so on. He marked and wrote down the following fractal formula:
$Z_1 = Z_0^2 + Z_0$
$Z_2 = Z_1^2 + Z_0$
$Z_3 = Z_2^2 + Z_0 \ldots$
If this sequence remains bounded, the complex number Z_0 belongs to the Mandelbrot set.
Fractal (Latin *fractus* – broken, smashed, split) is a geometric object that is similar to itself. Its main properties remain intact if we investigate the part of the fractal, similar (in one way or another) to the whole object.

even in Bach's and Mozart's music. That is because in these composers' work the acoustic model, or the fundamental structure, remains when compressed according to fractal principle. For example, the fundamental tonal scale of Bach's clavier inventions remains the same even after getting rid of 1/2, then 1/3, 1/4 and so on, numbers of tones (Hsü & Hsü 1991: 98; also see Lewin 1991). The self-repetition of the fractal macromodel in smaller levels can be seen in Schenker's theory as well, because the reductive experiments in the musical composition, the derivation of *Vordergrund, Hintergrund* and *Urlinie*, recall the principles of self-similarity.

I would like to discuss an example of the computer program MusiNum for composing music, which shows ways of carrying fractals over into tonal space. The computer code was created and presented by Lars Kindermann (presented in Kinderman 2006). In the beginning diatonic musical tones are matched with the binary numeral system: the tone *c* is matched with numeral combinations that have one symbol "one" (1) and the varying quantity of symbol "zero" (0), accordingly 1, 10, 100, 1000 etc.; numeral combinations of the tone *d* have two symbols "one" and the varying quantity of symbol "zero" (11, 101, 110, 1001 etc.); tone *e* has three symbols "one" and the varying quantity of symbol "zero" (111, 1011, 1101, 1110 etc.) and so on. Later, the binary codes are lined up into a row of decimal numbers in increasing order, and the musical tones are lined up matching their binary codes. Respectively, we get the transcription of data into a musical tone-sequence that creates the peculiarity of self-similarity. This is because every second, fourth, eighth, and so on, level's tone-sequence recalls its original model (see Figure 110).

Often composers themselves discuss the implications of fractals in musical composition. According to Bruno Degazio (born 1958) the piece *Roads to Chaos* (1986) was composed using fractal processes (Degazio 1986: 440). In his computer music piece *Profile* (1984) Charles Dodge used the computer to match pitch, rhythm, elements of amplitude to the 1/f noise algorithm. Dodge describes this work as a recursive structure that fills in time, analogical to the principle of filling space with fractals. This is because the melody for three parts (voices) was constructed as follows: each upper voice tone in the middle (second) voice is expanded to a phrase of a few tones. And then each tonal unit of this voice becomes the "seed" of the bottom (third) voice phrase (Dodge 1988: 11–4; see Figure 111).

Figure 110. Construction of fractal musical melody according to MusiNum (reproduced from Kinderman 2006)

c – 1 = 1, 10, 100, 1000 etc.
d – 2 = 11, 101, 110, 1001, 1010, 1100 etc.
e – 3 = 111, 1011, 1101, 1110 etc.
f – 4 = 1111, 10111, 11011, 11101 etc.
g – 5 = 11111, 101111, 110111, 111011, 111101, 111110 etc.
a – 6 = 111111, 1011111, 1101111, 1110111, 1111011, 1111101, 1111110 etc.
b – 7 = 1111111, 10111111, 11011111, 11101111, 11110111, 11111011, 11111101, 11111110 etc.

binary number	summing up of units	corresponding tone	selection of every 2nd	4th tone
1	1 (1)	*c*		
10	1 (1 + 0)	***c***	c	
11	2 (1 + 1)	*d*		
100	1 (1 + 0 + 0)	***c***	c	c
101	2 (1 + 0 + 1)	*d*		
110	2...	***d***	*d*	
111	3	*e*		
1000	1	***c***	c	c
1001	2	*d*		
1010	2	***d***	*d*	
1011	3	*e*		
1100	2	***d***	*d*	*d*
1101	3	*e*		
1110	3	***e***	*e*	
1111	4	*f*		
10000	1	***c***	c	c

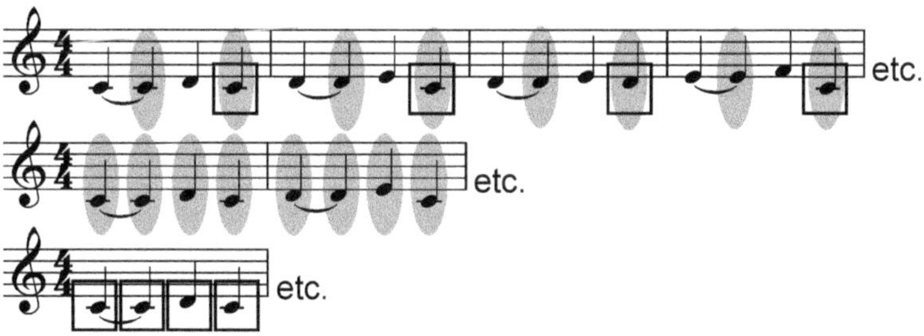

Figure 111. Dodge, Profile. *Principle of fractality in creating three-voice composition*

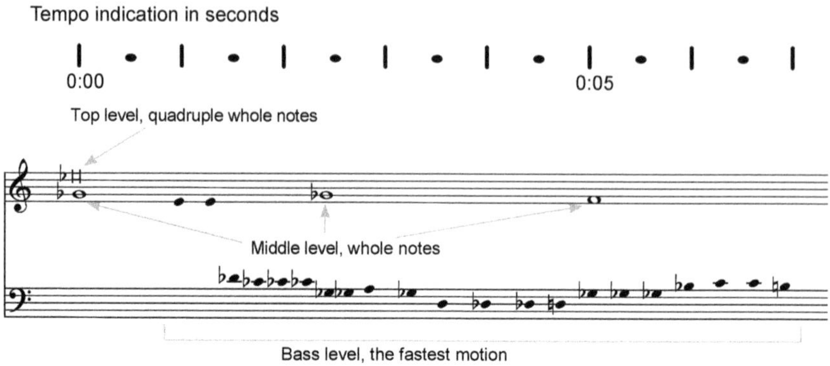

Figure 112. Johnson, Kientzy Loops. *Fractal arrangement of melody*

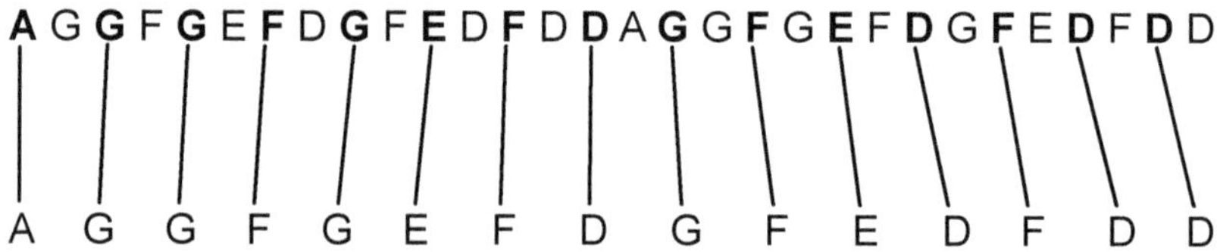

Gary Lee Nelson indicates that he used theories of fractals, chaos, artificial intelligence, quaternions, iterated function systems, and the L-system, as the sources for his work. In his microtone composition *Fractal Mountains* (1988–9), he applied a recursive division of musical time, pitch, and amplitude according to fractal algorithms. The composer merged this piece's form with the contours of fractal mountains, thus creating a microtone system, which divides an octave into 96 even intervals of 12.5 cents (Nelson 1993: 2 & 1996: 3).

Tom Johnson's composition *Kientzy Loops* (2000) illustrates how the principle of self-similarity can be found in tonal material. The perpetually playing phrase is heard as the same melody playing at a tempo that is two, three and more times slower (see Figure 112).

Šarūnas Nakas. *Ziqquratu* (1998)

In part two of this book, Nakas' composition *Ziqquratu* was presented as an instance of number squares transformed into musical material. However, the musical analysis may enrich this piece with the search for its fractal features. This can be investigated in the piano part, which is composed from symmetrical and asymmetrical combinations of sound clusters.

The composer claims that his asymmetrical tonal combinations were inspired by certain geometric forms, whose visual expression was carried over into the realm of tonal relationships. When I completed a graphic transcription of the asymmetrical clusters, I noticed the cluster pairs over a distance connect into symmetrical images. In order to compose the piano part, three interrelated symmetrical graphic models were used. Then Nakas divided them in half and "scattered" their parts throughout the musical composition. A graphic analysis of the asymmetrical piano clusters allowed me to equate the tonal dynamics of the piano part together with examples from fractal geometry. This is because in the united graphics of asymmetric sections we may see a visual similarity with a fractal, the contours of Koch snowflake (see Figures 113 & 114).[185] The resemblance to the Koch snowflake is rather imaginary, and, first of all, is based on visual generality. Therefore, this relationship can be better described by a quasi-fractal approach.

Figure 113. Nakas, Ziqquratu. *Graphics of a joined two assymmetrical clusters in visual comparison with a Koch snowflake*

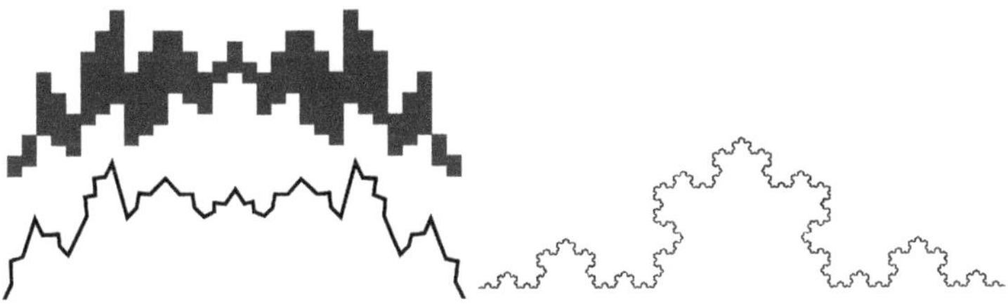

185 The Koch snowflake is a geometric figure, which in 1904 was discovered by the Swedish mathematician Helge von Koch. The construction of the Koch snowflake is a recursive process: 1) one begins with a black equilateral triangle of arbitrary size, 2) each side of the triangle is divided into three equal parts. A small triangle is attached to the middle part. Thus the derived star object has 12 sides, 3) further, each of those 12 sides is similarly divided along with the action of adding the small triangle. The creation of the Koch snowflake is made according to consistent steps – each triangle's side is changed into .

Figure 114. Nakas, Ziqquratu. *A pair of clusters, from m. 38 and m. 192*

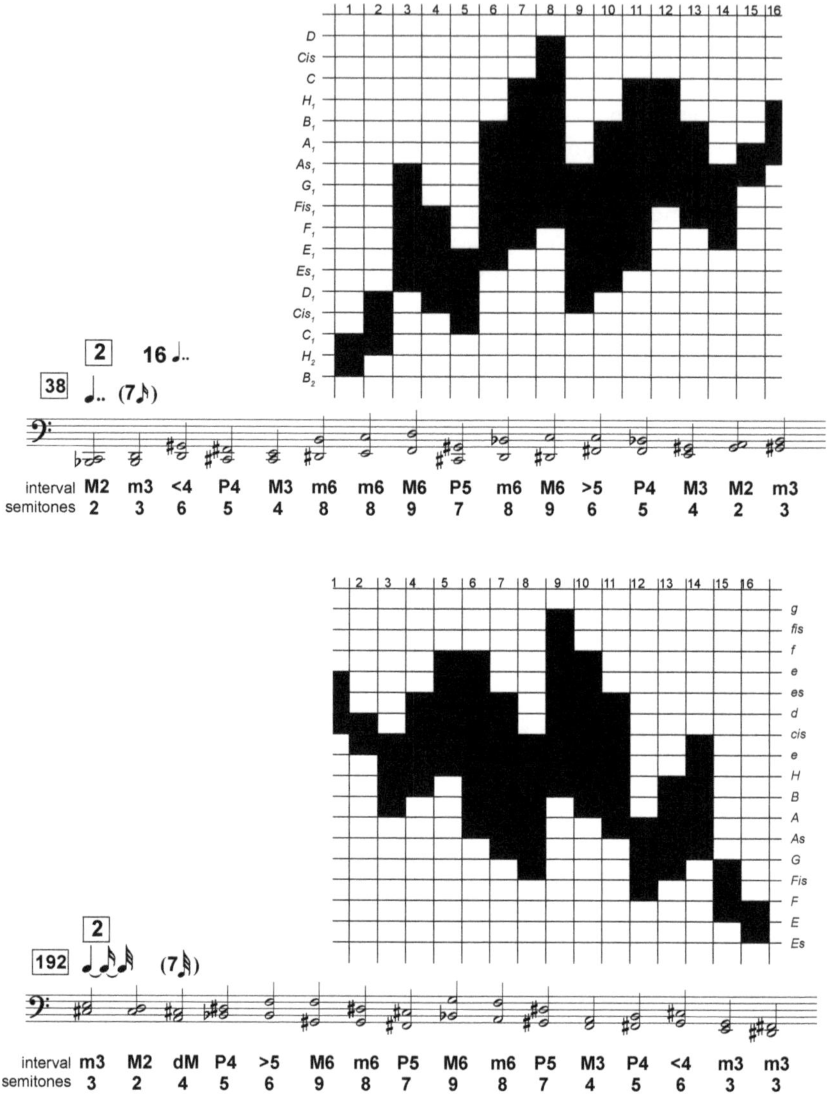

Vytautas V. Jurgutis. *Fractals* (1999)

Šarūnas Nakas' application of fractal geometry serves a rather abstract, or semantic, purpose. Whereas the Lithuanian composer Vytautas V. Jurgutis' (born 1976) composition for an instrumental ensemble, *Fractals* (1999), demonstrates a constructive employment of the fractal principle. The piece is based on the self-similarity of musical scales, which is general to all instrumental parts.[186] For example, the analysis of the cello part reveals the principle of the self-similar scale that manifests in the repetition and sequences of the four-tone motif in different levels (see Figure 115).

The melodic motion in the first piano part (marked in the score as *Keyboard I*) is constructed in the F-sharp minor scale, enriched with the diatonic rise of the fifths. The graph of the first piano part shows that the scale is accurately repeated in three levels: the original scale, or micromodel, *f-sharp – g-sharp – a – b – c-sharp – d – e – f-sharp* manifests itself on a larger level and so on (see Figure 116).

In the second piano part (marked as *Keyboard II*) the tonal sequence is created by manipulating the diatonic scale of F-sharp minor. The fractal principle of the scale expresses itself on four levels (see Figure 117).

The saxophone part is made up of a chain of seventh chords. The perpetual motion of chord arpeggios, every time starting from the next tone, forms a B-major scale on a larger plane (see Figure 118).

Figure 115. Jurgutis, Fractals. *Self-similarity principle in the sequences of the four-tone motif in different levels, the cello part, mm. 73–151*

186 Jurgutis' *Fractals* was composed for two piccolo flutes, saxophone, two pianos, violin, cello, marimba, electric and bass guitars.

Figure 116. Jurgutis, Fractals. *Self-similarity principle in an ascending scales of the first piano part, mm. 1–9*

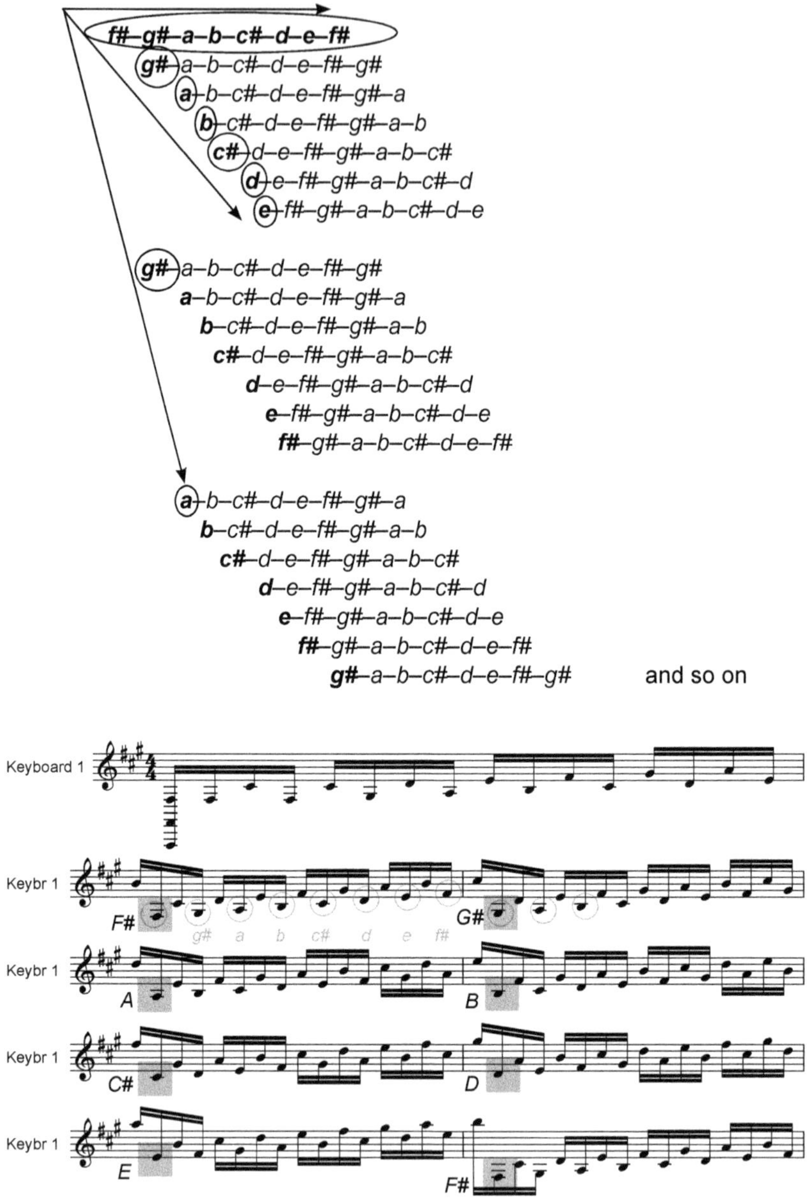

Figure 117. *Jurgutis,* Fractals. *Self-similarity principle in an ascending diatonic scale of the second piano part, mm. 71–102*

Figure 118. *Jurgutis,* Fractals. *Chain of seventh chords, forming a B-major scale, the saxophone part*

Charles Dodge. *A Fractal for Wiley Hitchcock* (1989)

My analysis of Dodge's composition for piano *A Fractal for Wiley Hitchcock* (1989) provides another instance of the implications of fractal rules. While composing this piece, fractals were used to organize the rhythmic material. This is the continual manipulation of the same rhythm formulas in original and diminutive forms.

From a structural point of view, the musical texture is made up of two dynamic plans. In the left hand part, the score is written down in prolonged rhythmic values. They are matched with small rhythmic units that are filled in by the right hand part. When I analyzed how the prolonged rhythmic values are brought into the composition, I applied the numerical equivalents to their rhythmic values and noticed a regularity of the durations. Therefore, I distinguished five groups of certain rhythmic values that are separated into five sections with a complex of rhythmic blocks. Thus the composition is made up of five sections of 8, 8, 5, 5 and 6 measures (see Figure 119).

While continuing the analysis of rhythmic blocks, I established that in the entire piano piece three groups of rhythmic formulas emerge. The first rhythmic formula group is connected with the third group as an original and as a retrograde, while the inner organization of the second group has typical mirror symmetry. Besides, each group consists of 4 to 5 rhythmic formulas that are diminutive derivations of their original model (see Figure 120). Recreating these observations in the first measures of the piece, I marked the rhythmic formulas with letters according to the order the composer used to present them in the score.

The analysis displayed that the rhythmic texture of the entire composition was modeled on a same chain made up of five rhythmic elements, a–b–c–a–c, and its combinatory manipulations. Each time in the chain one to four elements are changed, while consistently maintaining the number sequence 1–2–3–2–4. That is, the first time one member (rhythmic formula) of the chain is changed, the second time – two, the third – three, the fourth time – two, and the fifth time – four (see Figure 121).

The chains of rhythmic formulas used in the composition interact with a mathematical diminution – each new chain is a double division of the chain that came before. This creates the repetitious sequence of numbers 1 : 2 : 4 : 8. The image of rhythmic diminution in the first section is presented in Figure 122. In summary the entire composition of *A Fractal for Wiley Hitchcock* is based on the regular self-repetition of the rhythmic model (see Figure 123).

Figure 119. *Dodge,* A Fractal for Wiley Hitchcock. *Numerical reduction of rhythmic structure, showing the form of five sections*

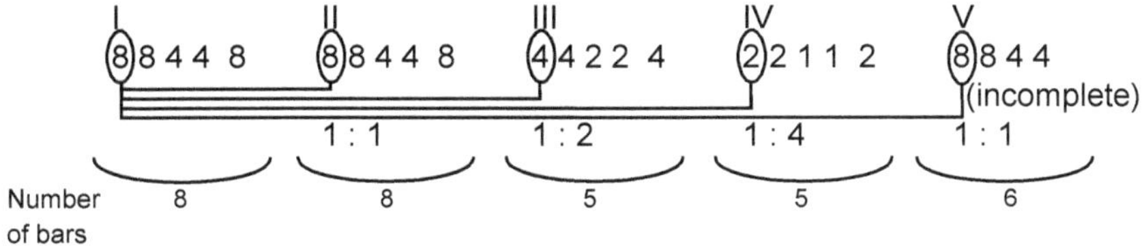

Figure 120. *Dodge,* A Fractal for Wiley Hitchcock. *Change of rhythmic formulas, mm. 1–4*

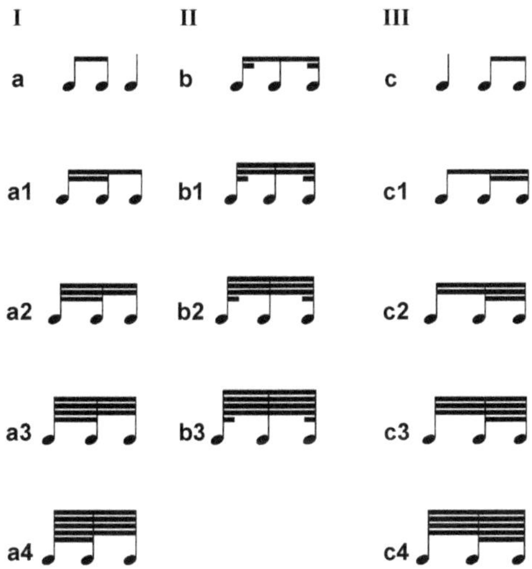

Figure 121. Dodge, A Fractal for Wiley Hitchcock. *Plan of rhythmic construction in letter symbols*

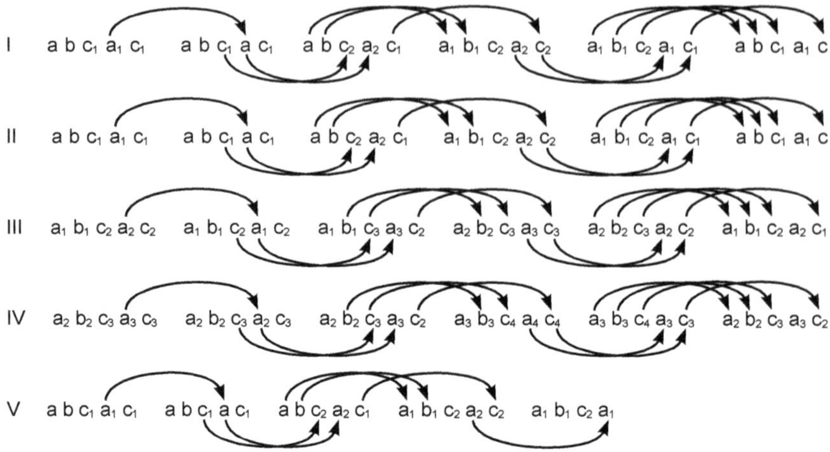

Figure 122. Dodge, A Fractal for Wiley Hitchcock. *Rhythmic design of the first section*

Figure 123. Dodge, A Fractal for Wiley Hitchcock. *Rhythmic pattern as an example of self-repetition*

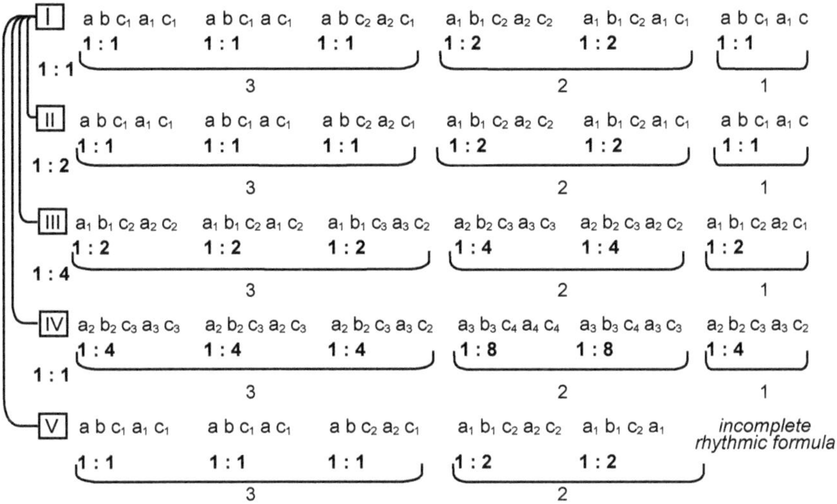

Afterword

The investigations of musical compositions considered in this book determine the original forms of mathematical influence on the art of sounds and on compositional possibilities. These investigations reveal the phenomenon of the renewal of mathematical traditions of earlier epochs, as well as the original trend representing the process of rendering contemporary music mathematically. However, the diversity presented here enables me to state that the original/universal system of composing, created by a composer,[187] manifests itself quite rarely. More often contemporary composers experiment with individual concepts within the framework of one composition. They individually choose traditionally defined ways, or elements of mathematical innovations, and adapt their chaotic combinations.

The following statement from Leibniz emphasizes the intuitive process of musical creation: "The pleasure we obtain from music comes from counting, but counting unconsciously. Music is nothing but unconscious arithmetic."[188] Meanwhile, the composing practice of contemporary music demonstrates a high degree of consciousness, internal logic, and a solidly created structure, or measures of proportions of certain parameters. These are usually determined in the advance sketches, pre-compositional plans and graphs.

Knowledge of the initial idea is often the only key to the correct analysis of a musical composition of the 20th–21st centuries, as well as an inseparable knowledge of other scientific fields that are closely related to the theories of advanced mathematics, fractals, chaos, etc. On the other hand, the practice of the computerized composing of music, which has recently accelerated, has opened the way for innovative analytical approaches. One could argue that it "shook" the fundamentals of traditionally established definitions of music, peculiarly initiating value-based transformations of the categories, such as creator and musical composition. First of all, in the modern world of information we encounter a *different* position of the composer. For example, Adam Alpern's comment (Alpern

---

187 Few individual examples of such original/personal systems are to be mentioned – Schönberg's dodecaphonic system, Webern's idea of the total series, Messiaen's non-reversible rhythms and system of interversions, in Lithuanian music: Balakauskas' dodecatonics, etc.

188 Latin *Musica est exercitium arithmeticae occultum nescientis se numerare animi.* From a letter to Christian Goldbach, April 27, 1712; quoted in Enrico Fubini, Bonnie J. Blackburn: *Music and Culture in Eighteenth-Century Europe: A Source Book*, Chicago: University of Chicago Press, 1994, p. 16.

1995: 1) on human's minimal participation in the creation of algorithmic music provokes a new definition of the composer's role within the context of the creative work. This is the gradual establishment of a composer-programmer model. The definition of a modern creator of music is focused on the person for whom mastery of computer programming becomes an inseparable part of his agenda. While paraphrasing David Cope, not all contemporary composers manage the complicated processes of computer programming, thus forming a category of *non-scientist composers.*[189] Considering aesthetic and functional points of view, I would like to note the increasing separation of a *scientist* and *non-scientist composer* or a *composer* and *art composer*[190] at the turn of the 20th to the 21st centuries. This concept brings to mind the division between professional composers and craftsmen in the musical panorama of the 18th–19th centuries. Additionally, the audience of computer music stimulates the development of a listener of a new kind, a listener who is able to *hear* the algorithm used in a composition. Is this prerequisite not in line with the concept of an educated listener in the Baroque epoch? A listener who was capable of recognizing rhetorical figures in music and deciphering the meanings hidden behind them?

The definition of the function of a computer, a special software, is becoming complicated as well. Is it merely a tool that generates sounds according to certain commands? Or is it an equivalent second "author", because the most complicated calculations, derivation of numerical formulas, and the actions of transcribing them into sounds, are entrusted to the computer? The mathematical algorithm applied by a computer determines the style, structure, even the final result. The composer participates in this process by presenting the software with the initial data – the "seed", then intervening in and modifying the process, and then making an assessment of the proposed result. This leads to the question about the authorship of the final result, a computer-generated musical composition. Should the music piece be attributed to the composer, the person, the initiator of the idea who gives specific instructions to a computer? Or should it be attributed to the developer, the creator of the computer software whose product carried out those procedures? Eventually, what is the dividing line? Where should it be drawn when transforming mechanical computer codes into a creative intention?

189 "I presume this concentration of these two categories will change over time, once more accessible information is made available to non-scientist composers." (quoted from Cope 2008: xiii)

190 Nierhaus in his book uses two conceptions: *composer* and *art composer* (more see Nierhaus 2009).

These considerations lead us to the term of *meta-composition*,[191] which focuses on what is more than an elementary or traditional musical composition. The concept of meta-composition has become especially convenient in defining the complexity of a musical composition in a computer space and in unifying the computer software encompassing the ideas of generating music and lots of possibilities as well as the final result. Though, from the methodological point of view, there is a lack of tools as, for example, naming the genre of a computer-generated composition. Even by making use of the same compositional models, composers can generate a lot of solutions and absolutely different results.

In the frame of the 20th–21st centuries, it becomes more and more difficult to apply the definitions and concepts of traditional music. Also, it is too early to speak about style in the computer music panorama that has been exploited over the last few decades. However, there is no doubt that the use of computer possibilities, the transformation of algorithmic processes into the space of music, and the manipulations with various mathematical phenomena considerably extend the spectrum of generating musical ideas. These processes have opened up immense layers of the implementation of the *musica mathematica* phenomenon.

191 *Greek μετά* – after, beyond, adjacent, self.

Bibliography

(Agrippa 1533) Heinrich Cornelius Agrippa: *De occulta philosophia libri tres*, Cologne.

(Alpern 1995) Adam Alpern: *Techniques for Algorithmic Composition of Music*, Hampshire College, http://citeseerx.ist.psu.edu/viewdoc/summary?doi=10.1.1.23.9364 [accessed 12 September 2016].

(Ames 1989) Charles Ames: The Markov Process as a Compositional Model: A Survey and Tutorial, in *Leonardo Music Journal*, Vol. 22, No. 2, pp. 175–187.

(Andreatta 2003) Moreno Andreatta: *Méthodes algébriques en musique et musicologie du XXe siècle: aspects théoriques, analytiques et compositionnels*, Thèse de doctorat, École des Hautes Etudes en Sciences Sociales.

(Andreatta 2004) Moreno Andreatta: On Group-Theoretical Methods Applied to Music: Some Compositional and Implementational Aspects, in *Perspectives in Mathematical and Computational Music Theory*, G. Mazzola, T. Noll & E. Lluis-Puebla (eds.), Osnabrück: Electronic Publishing Osnabrück, pp. 169–193.

(Andreatta 2007) Moreno Andreatta, and Carlos Agon: Structure and Symmetry in Iannis Xenakis Nomos Alpha for Cello Solo, in *Symmetry: Art and Science*, G. Lugosi & D. Nagy (eds.), 2–4. Actes du Colloque Form and Symmetry, Buenos Aires, 11–17 Novembre 2007, pp. 70–73.

(Andreatta 2011) Moreno Andreatta: Constructing and Formalizing Tiling Rhythmic Canons: A Historical Survey of a 'Mathemusical' Problem, in *Perspectives of New Music*, Vol. 49, No. 2 (Summer), pp. 33–64.

(Andrijauskas 1996) Antanas Andrijauskas: *Grožis ir menas. Estetika ir meno filosofijos idėjų istorija. Rytai–Vakarai* [Beauty and Art. History of Ideas of Aesthetics and Philosophy of Art: East-West], Vilnius: Vilniaus dailės akademijos leidykla.

(Apel 1970) Willi Apel, *Harvard Dictionary of Music*, 2nd ed., Cambridge, Massachusetts.

(Atlas 1987) Allan W. Atlas: Gematria, Marriage Numbers, and Golden Sections in Dufay's *Resvellies vous*, in *Acta Musicologica*, Vol. 59, Mai–August, pp. 111–126.

(Atlas 2003) Allan W. Atlas: Stealing a Kiss at the Golden Section: Pacing and Proportion in the Act I Love Duet of La Bohème, in *Acta Musicologica*, Vol. 2, pp. 269–291.

(Baroni et al. 1982) Mario Baroni, Rossella Brunetti, Laura Callegari, and Carlo Jacobini: A Grammar for Melody. Relationships between Melody and Harmony, in *Musical Grammars and Computer Analysis*, M. Baroni & L. Callegari (eds.), Proceedings of the conference, Modena, Italy, 4–6 October 1982 (Florence: Olschki, 1984), pp. 201–218.

(Barthelmes 1985) Barbara Barthelmes: Polyphonie der Proportionen. Zum Verhältnis von Architektur und Musik in der Renaissance, in *Musica*, Vol. 39, pp. 129–136.

(Bayreuther 2004) Rainer Bayreuther: Johannes Keplers musiktheoretisches Denken, in *Musiktheorie*, 19. Jahrgang, Heft 1, pp. 3–20.

(Birnbaum 2006) Jens Birnbaum: *Der Apollontempel von Didyma – Analyse einer pythagoreisch-platonischen Entwurfskonzeption*, Doctoral Dissertation, Technical University of Berlin, School of Architecture.

(Brauneiss 1995) Leopold Brauneiss: Proportionen und Zahlen in Mozarts Sonate facile, in *Musica*, 49. Jahrgang, 6 Hefte, pp. 86–90.

(Brinkman & Mesiti 1991) Alexander R. Brinkman, and Martha R. Mesiti: Graphic Modeling of Musical Structures, in *Computers in Music Research*, Vol. 3, pp. 1–42.

(Bruhn 1997) Siglind Bruhn: *Images and Ideas in Modern French Piano Music: The Extra-Musical Subtext in Piano Works by Ravel, Debussy, and Messiaen*, Stuyvesant, New York: PendragonPress.

(Bukofzer 1939/40) Manfred Bukofzer: Allegory in Baroque Music, in *Journal of the Warburg and Courtauld Institute*, Vol. 3, pp. 1–21.

(Burns 1994) Kristine H. Burns: *The History and Development of Algorithms in Music Composition, 1957–1993*, DA diss., UMI, Music: Ball State University.

(Cenova 2000) Валерия С. Ценова: *Числовые тайны музыки Софии Губайдулиной* [Numerical Secrets in the Music of Sofia Gubaidulina], Moscow: Moscow Conservatoire.

(Cholopov 1982) Юрий Н. Холопов: Изменяющееся и неизменимое в эволюции музыкального мышления [Change and Stability in the Development of Musical Thought], in *Проблемы традиций и новаторства в современной музыке* [Problems of Traditions and Innovations in the Contemporary Music], Moscow, pp. 52–104.

(Cholopov 1988) Юрий Н. Холопов: *Гармония. Теоретический курс* [Harmony: Theoretical Approach], Moscow: Muzyka.

(Clendinning 1995) Jane Clendinning: Review on Miguel A. Roig-Francoli *Harmonic and Formal Processes in Ligeti's Net-Structure Compositions*, in *Mu-*

sic *Theory Online*, Vol. 2, No. 5, http://www.mtosmt.org/issues/mto.96.2.5/mto.96.2.5.clendinning.html [accessed 12 September 2016].

(Contemporary Composers 1992) *Contemporary Composers*, B. Morton & P. Collins (eds.), Chicago and London: St. James Press.

(Conus 1986) Georgij E. Conus: Die metrotektonische Lösung des Problems der musikalischen Form, in *Musiktheorie*, 1. Jahrgang, Heft 1, pp. 83–88.

(Cope 2000) David Cope: *The Algorithmic Composer*, Madison, Wisconsin: A-R Editions.

(Cope 2008) David Cope: Preface, in *The OM Composer's Book*, Vol. 2, coll. Musique/Sciences, J. Bresson, A. Agon & G. Assayag (eds.), Paris: Editions DELATOUR FRANCE/Ircam-Centre Pompidou, pp. ix–xiv.

(Čiurlionis 2011) Mikalojus Konstantinas Čiurlionis: *Laiškai Sofijai* [Letters to Sofia], V. Landsbergis (ed.), Vilnius: Baltos lankos.

(Dahlhaus 1985) Carl Dahlhaus: Musik und Zahl. Zur Geschichtlichkeit eines metaphysischen Prinzips, in *Daidalos*, Vol. 17, pp. 18–25.

(Dammann 1967) Rolf Dammann: *Der Musikbegriff im deutschen Barock*, Cologne: Arno Volk Verlag.

(Daunoravičienė 2000) Gražina Daunoravičienė: Kompozicinės technikos algoritmai: Antroji simfonija ir „Dada concerto“[Algorithms of Compositional Technique: The Second Symphony and Dada Concerto], in *Osvaldas Balakauskas. Muzika ir mintys* [Osvaldas Balakauskas: Music and Ideas], R. Gaidamavičiūtė (ed.), Vilnius: Baltos lankos, pp. 71–122.

(Degazio 1986) Bruno Degazio: Musical Aspects of Fractal Geometry, in *Proceedings of the International Computer Music Conference*, P. Berg (ed.), San Francisco: Computer Music Association, pp. 435–442.

(Delahaye 2004) Jean-Paul Delahaye: La musique mathématique de Tom Johnson, in *Pour la Science*, No. 325 (Novembre), pp. 88–93.

(DeLio 1980) Thomas DeLio: Iannis Xenakis' *Nomos Alpha*: The Dialectics of Structure and Materials, in *Journal of Music Theory*, Vol. 24, No. 1 (Spring), pp. 63–95.

(Dikčiūtė 1993) Snieguolė Dikčiūtė, Violeta Bendinskaitė: Septyni tiltai į save [Seven Bridges to Yourself], in *Katalikų pasaulis* [Catholic World], Vilnius, No. 3 (March), pp. 8–9.

(Dodge & Bahn 1986) Charles Dodge, and Curtis R. Bahn: Musical Fractals: Mathematical Formulas Can Produce Musical as well as Graphic Fractals, in *Byte*, Vol. 11, No. 6 (June), pp. 185–196.

(Dodge 1988) Charles Dodge: Profile: A Musical Fractal, in *Computer Music Journal*, Vol. 12, No.3, pp. 10–14.

(Eberlein 1994) Dorothee Eberlein: Čiurlionis, Skrjabin und der osteuropäische Symbolismus, in *Vom Klang der Bilder*, K. von Maur (ed.), Stuttgart, pp. 340–345.

(Ehrler 2000) Hanno Ehrler: Mit Morzezeichen und Zahlencodes. Zu Gerhard Stäblers Komposition *O Muro*, in *Musik Texte*, Vol. 86–87 (November), pp. 13–17.

(Entress 1998) Matthias Entress: Review of Johnson's *The Chord Catalogue*, in *Berliner Morgenpost*, 24 November.

(Escot 1999) Pozzi Escot: *The Poetics of Simple Mathematics in Music*, Cambridge.

(Fedotov 1989) Владимир Михайлович Федотов: *Музыкальные основы творческого метода Чюрлениса* [Čiurlionis' Creative Methods: Musical Background], Saratov University.

(Frese 1998) Martha Ida Frese: *Wer war Sarastro, wer Tamino? Die Zauberflöte als Allegorie zur politischen Szene im josephinischen Wien*, Pähl: Verl. Hohe Warte.

(Gann 1987) Kyle Gann: 8178-Chord Wonder (Tom Johnson, Phill Niblock), in *Village Voice*, 14 April, Vol. 22, No. 15, p. 84.

(Gann 1995) Kyle Gann: *The Music of Conlon Nancarrow*, Cambridge: University Press.

(Gardner 1974) Martin Gardner: Mathematical Games: the Arts as Combinatorial Mathematics, in *Scientific American*, Vol. 231, No. 6, pp. 132–136.

(Gardner 1992) Martin Gardner: *Fractal Music, Hypercards and More… Mathematical Recreations from Scientific American Magazine*, New York: W. H. Freeman and company.

(Gerver & Lebedeva 1999) Лариса Гервер, and Анна Лебедева: Ars Combinatoria XVII–XVIII веков – прообраз музыкальной техники XX века [Ars Combinatoria in the 17th–18th centuries: A prototype of Musical Technique of the 20th century], in *Искусство XX века: Диалог эпох и поколений* [Art in the 20th century: Dialogues between Epochs and Generations], Vol. 2, B. S. Getselev & T. Sidneva (eds.), Nizhny Novgorod, pp. 30–50.

(Gratzer 1993) Wolfgang Gratzer: *Zur „wunderlichen Mystik" Alban Bergs*, Wien: Böhlau.

(Gravenhorst 1995) Tobias Gravenhorst: *Proportion und Allegorie in der Musik des Hochbarock. Untersuchungen zur Zahlenmystik des 17. Jahrhunderts mit beigefügtem Lexikon*, Frankfurt am Main: Peter Lang, 1995.

(Griffiths 1981) Paul Griffiths: *Cage*, New York: Oxford University Press.

(Griffiths/Ligeti 2001) Paul Griffiths: Ligeti, György (Sándor), in *The New Grove Dictionary for Music and Musicians*, S. Sadie & J. Tyrrell (eds.), London: Macmillan, Vol. XIV, pp. 690–696.

(Griffiths/Messiaen 2001) Paul Griffiths: Messiaen, Olivier (Eugene Prosper Charles), in *The New Grove Dictionary for Music and Musicians*, S. Sadie & J. Tyrrell (eds.), London: Macmillan, Vol. XVI, pp. 491–504.

(Griffiths 2012) Paul Griffiths: Cryptography, in *The Oxford Companion to Music,* A. Latham (ed.), *Oxford Music Online*, Oxford University Press, http://www.oxfordmusiconline.com/subscriber/article/opr/t114/e1750 [accessed 12 September 2016].

(Gumauskaitė 2000) Vida Gumauskaitė: *Struktūralizmo apmatai. Sąvokos, metodas, filosofija* [Fundamental Structuralism: Concepts, Methods and Philosophy], Vilnius.

(Hall 1928) Manly P. Hall: *The Secret teachings of All Ages*, Chapter 'The Pythagorean Theory of Music and Color', pp. 81–84, http://www.sacred-texts.com/eso/sta/sta19.htm [accessed 12 September 2016].

(Hammel 2015) Bill Hammel: *An Essay on Patterns in Musical Composition Transformations, Mathematical Groups, and the Nature of Musical Substance*, 1997–2011, http://graham.main.nc.us/~bhammel/MUSIC/compose.html [accessed 12 September 2016].

(Heher 1992) Erich Heher: *Beziehungen zwischen Mathematik und Musiktheorie im wandel der Zeiten*, Diplomarbeit für die Studienrichtung Musikerziehung, Hochschule für Musik und darstellende Kunst, Wien.

(Horner & Goldberg 1991) Andrew Horner, and David E. Goldberg: Genetic Algorithms and Computer-Assisted Music Composition, in *Proceedings of the 1991 International Computer Music Conference*, B. Alphonce & B. Pennycook (eds.), San Francisco: International Computer Music Association, pp. 479–482.

(Howat 1983) Roy Howat: *Debussy in Proportion*, Cambridge.

(Hsü & Hsü 1991) Kenneth Hsü, and Andrew Hsü: Self-Similarity in Music, in *Computing in Musicology*, Vol. 7, pp. 98–100.

(Huntley 1970) Herbert Edwin Huntley: *The Divine Proportion*, Dover Publications.

(Husmann 1961) Heinrich Husmann: *Grundlagen der Antiken und Orientalischen Musikkultur*, Berlin: Walter de Gruyter.

(Iamblichus 1818) Iamblichus: *Life of Pythagoras, or, Pythagoric Life. Accompanied by Fragments of the Ethical Writings of Certain Pythagoreans in the Doric Dialect*, Th. Taylor (transl.), London: J. M. Watkins.

(Janeliauskas 2010) Rimantas Janeliauskas: *Neatpažinti Mikalojaus Konstantino Čiurlionio muzikos ciklai* [Unrecognized Music Cycles by Čiurlionis], Vilnius: Lithuanian Composers' Union, Lithuanian Academy of Music and Theatre.

(Jankauskienė 2001) Inga Jankauskienė: Broniaus Kutavičiaus kompozicijos – „svetimo" raiška [The Expression of "Alien" in the Compositions by Bronius Kutavičius], in *Menotyra* [Art Research Journal], No. 3, pp. 68–77.

(Järveläinen 2000) Hanna Järveläinen: Algorithmic Musical Composition, in *Art@Science*, Spring, https://engineering.purdue.edu/ece477/Archive/2006/Fall/F06-Grp01/files/alco.pdf [accessed 12 September 2016].

(Johnson *Editions 75*) Tom Johnson: *Editions 75 Catalogue*, http://www.editions75.com/English/catalogenglish.html [accessed 12 September 2016].

(Johnson 1989) Tom Johnson: *The Voice of New Music*, Eindhoven: Het Apollohuis, http://www.editions75.com/Books/TheVoiceOfNewMusic.PDF [accessed 12 September 2016].

(Johnson 2001) Tom Johnson: *Musique, mathématiques et philosophie*, Seminaire Entre Temps, IRCAM, 2001 01 13, http://www.entretemps.asso.fr/Seminaire/Johnson/index.html [accessed 12 September 2016].

(Johnson 2011) Tom Johnson: Tiling in My Music, in *Perspectives of New Music*, Vol. 49, No. 2 (Summer), pp. 9–21.

(Jones 2002) Evan Jones: An Acoustic Analysis of Col Legno Articulation in Iannis Xenakis' *Nomos Alpha*, in *Computer Music Journal*, Vol. 26, No. 1, pp. 73–86.

(Jones 2004) Kevin Jones: Breaking Elgar's Enigmatic Code, in *New Scientist*, 25 December, Issue 104 (2479/80), p. 56.

(Kee 1984) Kee, Piet: Maß und Zahl in Passacaglia und Ciacona – Astronomie in Buxtehudes Passacaglia, in *Ars Organi*, 32. Jahrgang, Heft 4, pp. 232–241.

(Kee 1988) Piet Kee: Zahl und Symbolik in Passacaglia und Ciacona, in *Musik und Kirche*, No. 58, pp. 231–255.

(Kepler 1619) Johannes Kepler: *Harmonices mundi*, Linz: J. Planck.

(Kindermann 2006) Lars Kindermann: *MusiNum – The Music in the Numbers*, 1995–2006, http://reglos.de/musinum/ [accessed 12 September 2016].

(Kirk 2001) Kenneth Patrick Kirk: *The Golden Ratio in Chopin's Preludes, Opus 28*, Doctoral Dissertation, UMI.

(Kirnberger 1757) Johann Philip Kirnberger: *Der allezeit fertige Polonoisen- und Menuettencomponist*, Berlin: G. L. Winter.

(Klotz 2006) Sebastian Klotz: *Kombinatorik und die Verbindungskunste der Zeichen in der Musik zwischen 1630 und 1780*, Berlin: Akademie Verlag.

(Korrick 2003) Leslie Korrick: Lamazzo's *Trattato... della pittura* and Galilei's *Fronimo*: Picturing Music and Sounding Images in 1584, in *Art and Music in the Early Modern Period: Essays in Honor of Franca Trinchieri Camiz*, K. A. McIver (ed.), Ashgate Publishing, pp. 193–215.

(Kramer 1973) Jonathan Kramer: The Fibonacci Series in 20th Century Music, in *Journal of Music Theory*, Vol. 17, No. 1, pp. 110–149.

(Kučinskas 2004) Darius Kučinskas: *M. K. Čiurlionio fortepijoninės muzikos tekstas (genezės aspektas)* [The Musical Text of Piano Works by M. K. Čiurlionis. Aspects of Genesis], Kaunas: Technologija.

(Kuhnau 1700/1973) Johann Kuhnau: *Musicalische Vorstellung einiger Biblischer Historien*, Leipzig, W. Reich (ed.), Leipzig: Edition Peters.

(Landsbergis 1976) Vytautas Landsbergis: *Čiurlionio dailė* [Čiurlionis' Art], Vilnius: Vaga.

(Lerdahl 2001) Fred Lerdahl: *Tonal Pitch Space*, Oxford University Press.

(Lévy 2011) Fabien Lévy: Three Uses of Vuza Canons, in *Perspectives of New Music*, Vol. 49, No. 2 (Summer), pp. 23–31.

(Lewin 1991) Roger Lewin: The Fractal Structure of Music, in *New Scientist*, Vol. 130: 19 (May).

(Ligeti 1988) György Ligeti: On My *Etudes* for Piano, On My Piano Concerto, in *Sonus*, Vol. 9, No. 1, pp. 3–14.

(Lobanova 1994) Марина Лобанова: *Западноевропейское музыкальное барокко: проблемы эстетики и поэтики* [Western European Baroque Music: Issues of Aesthetics and Poetics], Moscow.

(Losev 1963) Алексей Федорвич Лосев: *История античной эстетики* [History of Ancient Aesthetics], Vol. 1 'Early Classicism', Moscow.

(Losev 1988) Алексей Федорвич Лосев: *История античной эстетики* [History of Ancient Aesthetics], Vol. 1–2 'The Last Epochs', Moscow.

(Losev 1999) Алексей Федорвич Лосев: *Самое само* [The Thing Itself], Moscow: Eksmo-press.

(Luchese 1988) Diane Luchese: Levels of infrastructure in Ligeti's *Volumina*, in *Sonus*, Vol. 9, No. 1 (Fall), pp. 38–58.

(Madden 1999) Charles Madden: *Fractals in Music: Introductory Mathematics for Musical Analysis*, High Art Press.

(Madden 2005) Charles Madden: *Fib and Phi in Music: The Golden Proportion in Musical Form*, High Art Press.

(Mason & Saffle 1994) Stephanie Mason, and Michael Saffle: L-systems, Melodies and Musical Structure, in *Leonardo Music Journal*, Vol. 4, pp. 31–38.

(Mason et al. 1988) Stephanie Mason, Michael Saffle, and Hannes Schütz: Musik und Mathematik, in *Musik in Geschichte und Gegenwart*, Vol. VI, pp. 790–799.

(Mathematics and Music 2002) *Mathematics and Music: A Diderot Mathematical Forum*, G. Assayag, H. G. Feichtinger & J. F. Rodrigues (eds.), Springer.

(Mattheson 1740) Johann Mattheson: *Grundlage einer Ehren-Pforte*, Hamburg.

(Maurer 1999) John A. Maurer: *A Brief History of Algorithmic Composition*, Stanford University, ccrma.stanford.edu/~blackrse/algorithm.html [accessed 12 September 2016].

(Mažeikis 1998) Gintautas Mažeikis: *Renesanso simbolinis mąstymas* [Symbolic Thinking of the Renaissance], Šiauliai: Šiauliai University.

(Mažulis 2001) Rytis Mažulis: Menzūrinių kanonų technika ir jos perspektyvos šiuolaikinėje muzikoje [The Technique of Mensural Canon and its Application Possibilities in Contemporary Music], in *Lietuvos muzikologija* [Lithuanian Musicology Journal], Vilnius, No. 2, pp. 60–72.

(Mello 1995) Chico Mello: Precision and Anarchy in Feldman's Work, in *Magazine of the Instituto Superior de Musica of the Universidad Nacional del Litoral*, No. 4 (December), http://www.cnvill.net/mfmello.htm [accessed 12 September 2016].

(Mikėnaitė 2000) Rima Mikėnaitė: Osvaldo Balakausko kompozicinės technikos principai [Principles of Osvaldas Balakauskas' Composing Technique], in *Osvaldas Balakauskas. Muzika ir mintys* [Osvaldas Balakauskas: Music and Ideas], R. Gaidamavičiūtė (ed.), Vilnius: Baltos lankos, pp. 11–18.

(Moritz 2002) Albrecht Moritz: *Stockhausen. Tierkreis (Zodiac). Trio version (1975/83)*, http://home.earthlink.net/~almoritz/tierkreistrioversion.htm [accessed 12 September 2016].

(Morley 1597) Thomas Morley: *Plaine and Easie Introduction to Practicall Musicke*, London: Peter Short.

(Musik: Geschichte 1954) *Musik: Geschichte ihrer Deutung*, Orbis Academicus, Problem-geschichten der Wissenschaft in Dokumenten und Darstellungen, H. Pfrogner (ed.), München: Verlag Karl Alber Freiburg.

(Nakas 2002) Šarūnas Nakas: *Maurojantis tvanas apačioje ir siautėjanti orgija viršuje* [The Roaring Deluge and Storming Bacchanalia], manuscript, Vilnius, 21 August.

(Nelson 1993) Gary Lee Nelson: Sonomorphs: an Application of Genetic Algorithms to the Growth and Development of Musical Organisms, in *Proceedings of the Fourth Biennial Arts and Technology Symposium*, New London, Connecticut: Connecticut College, pp. 155–169.

(Nelson 1994) Gary Lee Nelson: *Wind, Sand and Sea Voyages: an Application of Granular Synthesis and Chaos to Musical Composition*, Oberlin, Conservatory of Music, USA, http://timara.con.oberlin.edu/gnelson/PapersPDF/gola.pdf [accessed 12 September 2016].

(Nelson 1996) Gary Lee Nelson: Real Time Transformation of Musical Material with Fractal Algorithms, in *Computers and Mathematics with Applications*, Vol. 32: 1, pp. 109–116.

(Nierhaus 2009) Gerhard Nierhaus: *Algorithmic Composition. Paradigms of Automated Music Generation*, Wien: Springer-Verlag.

(Peck 2003) Robert W. Peck: Towards an Interpretation of Xenakis' *Nomos alpha*, in *Perspectives of New Music*, January, pp. 106–158.

(Povilionienė 2003) Rima Povilionienė: *Konstruktyvinės ir simbolinės skaičiaus funkcijos XX a. muzikoje* [The Role of Numerical Construction and Symbols in 20th Century Music], Master's thesis, Vilnius: Lithuanian Academy of Music and Theatre.

(Povilionienė 2004) Rima Povilionienė: Technologinis ir simbolinis skaitmens konstruktyvizmas George'o Crumbo *Black Angels* ir György Ligeti *Etudes* [Technical and Symbolic Numerical Construction in George Crumb's *Black Angels* and György Ligeti's *Etudes*], in *Tiltai* [Bridges], No. 21, Klaipėda, pp. 137–148.

(Povilionienė 2006) Rima Povilionienė: Music Composition as a Numerical Construction. The Projections of Magic Number Quadrates and Dimensions of Ziggurat Pyramids in the Composition *Ziqquratu* (1998) by Šarūnas Nakas, in *Musical Work: Boundaries and Interpretations*, A. Žiūraitytė (ed.), Vilnius, pp. 278–291.

(Povilionienė 2013) Rima Povilionienė: Čiurlionio autografo ir dedikacijos tyrimai. Fuga *in b*: struktūrinė ir semantinė analizė [Search of Čiurlionis Signature and Dedication. Fugue in B-flat: Structural and Semantic Analysis], in *Mikalojus Konstantinas Čiurlionis (1875–1911): jo laikas ir mūsų laikas* [Mikalojus Konstantinas Čiurlionis (1875–1911): His Time and Our Time], G. Daunoravičienė & R. Povilionienė (eds.), Vilnius: Lithuanian Academy of Music and Theatre, pp. 268–301.

(Powel 1979) Newman W. Powel. Fibonacci and the Gold Mean: Rumbas, Rabbits and Rondeaux, in *Journal of Music Theory*, No. 23, pp. 227–273.

(Pritchett 2001) James Pritchett: Cage, John, in *The New Grove Dictionary for Music and Musicians*, S. Sadie & J. Tyrrell (eds.), London: Macmillan, Vol. XIV, pp. 690–696.

(Prusinkiewicz 1986) Przemysław Prusinkiewicz: Score Generation with L-Systems, in *Proceedings of the International Computer Music Conference*, P. Berg (ed.), San Francisco: Computer Music Association, pp. 455–457.

(Puckette 2006) Miller Puckette: Preface, in *The OM Composer's Book*, Vol. 1, coll. Musique/Sciences, A. Agon, G. Assayag & J. Bresson (eds.), Paris: Editions DELATOUR FRANCE/Ircam-Centre Pompidou, pp. ix–xiv.

(Riepel 1757) Joseph Riepel: *Gründliche Erklärung der Tonordnung insbesondere*, Frankfurt, Leipzig.

(Rockwell 1999) John Rockwell: *Morton Feldman (and Crippled Symmetry)*, Notes to CD recording of *Crippled Symmetry* by the California EAR Unit on the *Bridge Records* label (BRIDGE 9092A/B), http://www.cnvill.net/mfrockwl.htm [accessed 12 September 2016].

(Rothstein 1995) Edward Rothstein: *Emblems of Mind: the Inner Life of Music and Mathematics*, New York.

(Rozenov 1982) Эмилий Карлович Розенов: *Статьи о музыке* [Articles on Music], Moscow.

(Ryschawy & Stoll 1988) Hans Ryschawy, and Rolf W. Stoll: Die Bedeutung der Zahl in Dufays Kompositionsart: *Nuper rosarum flores,* in *Musik-Konzepte*, Heft 60 (April) 'Guillaume Dufay', pp. 3–73.

(Saleh Pascha 2004) Khaled Saleh Pascha: *Gefrorene Musik. Das Verhältnis von Architektur und Musik in der ästhetischen Theorie*, Doctoral Dissertation, Technical University of Berlin, School of Architecture.

(Sams 1971) Eric Sams: Brahms and his Musical Love-Letters, in *The Musical Times*, 1971, Vol. 112, No. 1538, pp. 329–330.

(Sams 2016) Eric Sams: Cryptography, musical, in *Grove Music Online. Oxford Music Online*, Oxford University Press, http://www.oxfordmusiconline.com/subscriber/article/grove/music/06915 [accessed 12 September 2016].

(Saunders 1996) James Saunders: *The Final Frontier: The Development of Polytempo in the Music of Conlon Nancarrow*, Master's Thesis, Royal Northern College of Music.

(Sautoy 2003) Marcus du Sautoy: *The Music of the Primes*, New York: Harper Collins.

(Schaller 1997) Erika Schaller: *Klang und Zahl. Luigi Nono: serielles Komponieren zwischen 1955 und 1959*, Saarbrücken: Pfau.

(Schouest 2000) Scott J. Schouest: *Symbolic Composition: Messiaen and the Roman Catholic Faith*, 11th Annual Conference Music Theory Midwest, Lawrence University, Appleton, WI, 19–21 May 2000, http://www.wmich.edu/mus-theo/mtmw/mtmw00/00_abs_fam.html [accessed 12 September 2016].

(Schütz 1997) Hannes Schütz: Wiedergeburt der Ars subtilior? Eine Analyse von György Ligetis Klavieretüde Nr. 2 *Cordes Vides*, in *Die Musikforsching*, Heft 2 (April–Juni), pp. 205–214.

(Shenton 2008) Andrew Shenton: *Olivier Messiaen's System of Signs. Notes Towards Understanding His Music*, Ashgate Publishing Ltd.

(Solomon 1973) Larry Solomon: New Symmetric Transformations, in *Perspectives of New Music*, Spring–Summer, pp. 257–264.

(Source Readings 1965) *Source Readings in Music History. The Reinaissance*, O. Strunk (ed.), New York: Princeton University.

(Soussidko 2002) Irina Soussidko: Metrotektonik und Goldener Schnitt. Debussy 24 Preludes für Klavier, in *Musik & Ästhetik*, 6. Jahrgang, Heft 24, pp. 5–19.

(Stechow 1953) Wolfgang Stechow: Problems of Structure in Some Relations between the Visual Arts and Music, in *The Journal of Aesthetics and Art Criticism*, Vol. 11, No. 4, pp. 324–333.

(Strohmayer 2001) Wolfgang Strohmayer: Alberti Leon Battista. Schönheitsbegriff und traditionelle Entwurfsgrundlagen, in *Archiv für Musikwissenschaft*, 58. Jahrgang, pp. 231–260.

(Supper 2001) Martin Supper: A Few Remarks on Algorithmic Composition, in *Computer Music Journal*, Vol. 25, No. 1, pp. 48–53.

(Takenouchi 1987) Aleksei Takenouchi: *Numbers and Proportions in George Crumb's Solo Piano Compositions*, Doctoral Dissertation, Northwestern University.

(Tarasevich 1992) Николай Иванович Тарасевич: Число и его значение в музыкальной композиции Возрождения (на примере мессы «Grecorum» Я. Обрехта) [Number and its Meaning in the Musical Compositions of Renaissance. An Example of Missa Grecorum by Obrecht], in *Методы изучения старинной музыки* [Research Methods of Medieval-Renaissance Music], G.V. Grigoryeva, T.N. Dubravskaya & L.N. Loginova (eds.), Moscow, pp. 43–55.

(Tatlow & Griffiths 2016) Ruth Tatlow, and Paul Griffiths: Numbers and music, in *Grove Music Online. Oxford Music Online*, Oxford University Press, http://www.oxfordmusiconline.com/subscriber/article/grove/music/44483 [accessed 12 September 2016].

(Tatlow 1991) Ruth Tatlow: *Bach and the Riddle of the Number Alphabet*, Cambridge University Press.

(Tatlow 2006) Ruth Tatlow: The Use and Abuse of Fibonacci Numbers and the Golden Section in Musicology Today, in *Understanding Bach*, Vol. 1, pp. 69–85.

(Tatom 2000) Marianne Tatom: *Sacred Mirrors: The Palindrome and the Ikon in John Tavener's The Protecting Veil*, 11th Annual Conference Music Theory Midwest, Lawrence University, Appleton, WI, 19–21 May 2000, http://www.

wmich.edu/mus-theo/mtmw/mtmw00/00_abs_fam.html [accessed 12 September 2016].

(Theory of Contemporary Music 2005) *Теория современной композиции* [Theory of Contemporary Music Composition], V. Cenova (ed.), Moscow.

(Terzi 1670) Francesco Lana de Terzi: *Prodromo all'Arte Maestra*, Brescia.

(Thomas 2000) Margaret E. Thomas: Nancarrow's Canons: Projections of Temporal and Formal Structures, in *Perspectives of New Music*, Vol. 38, No. 2 (Summer), pp. 106–133.

(Towsey et al. 2001) Michael Towsey, Andrew Brown, Susan Wright, and Joachim Diederich: Towards Melodic Extension Using Genetic Algorithms, in *Journal of Educational Technology and Society*, Vol. 4 (2), pp. 54–65.

(Venckus 2000) Antanas Venckus: M. K. Čiurlionio Fuga Op. 34 [Čiurlionis Fugue Op. 34], in *M. K. Čiurlionis ir lietuvių muzikinės kultūros raida* [Čiurlionis and the Development of Lithuanian Musical Culture], Vilnius: Lithuanian Academy of Music and Theatre, pp. 175–253.

(Voss & Clarke 1978) Richard F. Voss, and John Clarke: 1/f Noise in Music, in *The Journal of the Acoustical Society of America*, Vol. 63: 1, pp. 258–263.

(Wager 1998) Gregg Wager: *Symbolism as a Compositional Method in the Works of Karlheinz Stockhausen*, College Park, Maryland: G. Wager.

(Walther 1732) Johann Gottfried Walther: *Musikalisches Lexikon oder musikalische Bibliothek*, Leipzig: W. Deer.

(Warner Marien & Fleming 2005) Mary Warner Marien, and William Fleming: *Arts and Ideas*, Belmont: Thomson/Wadsworth.

(Warren 1973) Charles Warren: Brunelleschi's Dome and Dufay's Motet, in *The Musical Quarterly*, Vol. 59, pp. 92–105.

(Wilkins 1694). John Wilkins, *Mercury, or the Secret and Swift Messenger*, 2nd ed., London, also see the 1st edition, 1641: http://www.light-of-truth.com/Secret_Messenger/secret.html [accessed 12 September 2016].

(Wille 1982) Rudolf Wille: Symmetrien in der Musik, in *Neue Zeitschrift für Musik*, Vol. 143, Heft 12, pp. 12–19.

(Woodruff 2006) Eliot Ghofur Woodruff: Metrical Phase Shifts in Stravinsky's *The Rite of Spring*, in *Music Theory Online*, Vol. 12, No. 1 (February), http://www.mtosmt.org/issues/mto.06.12.1/mto.06.12.1.woodruff_frames.html [accessed 12 September 2016].

(Woolf 2002) Jonathan Woolf: Morton Feldman (1923–1987). Crippled Symmetry (1983), in *MusicWeb-International*, http://www.musicweb-international.com/classRev/2002/Dec02/MortonFELDMAN.htm [accessed 12 September 2016].

(Wright 1994) Craig Wright: Dufay's *Nuper rosarum flores*, King Solomon's Temple, and the Veneration of the Virgin, in *Journal of the American Musicological Society*, Vol. 47, No. 3 (Fall), pp. 395–441.

(Wurm 1984) Karl Wurm: Christus Kosmokrator. Ein hermeneutischer Versuch zu D. Buxtehudes Passacaglia in d BuxWv 161 und zu J. S. Bachs Präludium und Fugue C-dur BWV 547, in *Musik und Kirche*, No. 54, pp. 263–271.

(Xenakis 1992) Iannis Xenakis: *Formalized Music. Thought and Mathematics in Composition*, Harmonologia Series No. 6, Stuyvesant, NY: Pendragon Press.

(Zarlino 1562). Gioseffo Zarlino: *Le istitutioni harmoniche*, Venetia: F. Senese.

Supplementary Literature on the Subject of this Book

Eytan Agmon: A Mathematical Model of the Diatonic System, in *Journal of Music Theory*, Vol. 1, 1989, pp. 1–25.

Carlos Agon, Moreno Andreatta: Modeling and Implementing Tiling Rhythmic Canons in the OpenMusic Visual Programming Language, in *Perspectives of New Music*, Vol. 49, No. 2 (Summer), 2011, pp. 66–91.

Emmanuel Amiot: Structures, Algorithms, and Algebraic Tools for Rhythmic Canons, in *Perspectives of New Music*, Vol. 49, No. 2 (Summer), 2011, pp. 93–142.

Moreno Andreatta, Carlos Agon: Guest Editor's Foreword, Special Issue: Tiling Problems in Music, in *Journal of Mathematics and Music*, Vol. 3, No. 2 (July), 2009, pp. 63–70.

Linda M. Arsenault: Iannis Xenakis' *Achorripsis*: The Matrix Game, in *Computer Music Journal*, Vol. 26, No. 1, 2002, pp. 58–72.

Anthony Ashton: *Harmonograph. A Visual Guide to the Mathematics of Music*, New York: Walker & Company, 2003.

Markus Bandur: The Composition of Meaning: Construction and Semantics in Karlheinz Stockhausen's 'Luzifer-Gruss vom Samstag aus Licht', in *Perspectives of New Music*, Vol. 37, No. 1 (Winter), 1999, pp. 157–178.

Andre Barbera: The Consonant Eleventh and the Expansion of the Musical Tetractys: a study of Ancient Pythagoreanism, in *Journal of Music Theory*, Vol. 28, 1984, pp. 191–223.

Ernst Bindel: *Die Zahlengrundlagen der Musik im Wandel der Zeiten*, I. Teil, Stuttgart, 1950.

Agnes Cäcilie Bohnert: *Die arca musarithmica Athanasius Kirchers*, Berlin: Mensch & Buch Verlag, 2010.

Leopold Brauneiss: *Zahlen zwischen Struktur und Bedeutung*, Frankfurt am Main: Peter Lang, 1997.

Siglind Bruhn: *Musikalische Symbolik in Olivier Messiaens Weichnachtsvignetten. Hermeneutisch-analytiswche Untersuchungen zu den Vingt regards sur l'Enfant-Jesus*, Publikationen des Instituts für Musikanalytik Wien, Band 4, Frankfurt am Main, Bern, New York, Paris: Peter Lang, 1997.

Anne Maria Busse Berger: Musical Proportions and Artihmetics in the Late Middle Ages and the Renaissance, in *Musica Disciplina*, Vol. 44, 1990, pp. 89–118.

Juri Cholopov: Die Spiegelsymmetrie in Anton Weberns Variationen fur Klavier Op. 27, in *Archiv für Musikwissenschaft*, 30. Jahrgang, Heft 1, 1973, pp. 26–43.

David Clark Little: Composing with Chaos. Applications of a New Science for Music, in *ISEA'95 Proceedings. The Sixth International Symposium on Electronic Art*, Montreal, 1995, pp. 170–176.

Hermann Conen: *Formel-Komposition. Zu Karlheinz Stockhausens musik der siebziger Jahre*, Kölner Schriften zur Neuen Musik 1, J. Fritsch & D. Kämper (eds.), Mainz: Schott's Söhne, 1991.

Henry Cowell: *New Musical Resources*, New York: Something Else Press, 1969.

John D. Cuciurean: Review of Mark Lindley and Ronald Turner-Smith *Mathematical Models of Musical Scales: a New Approach*, in *Music Theory Online*, Vol. 1, No. 4 (July), 1995, http://www.mtosmt.org/issues/mto.95.1.4/mto.95.1.4.cuciurean.rev [accessed 12 September 2016].

Rolf Dammann: Zur Musiklehre des Andreas Werckmeister, in *Archiv für Musikwissenchaft*, Vol. 11, 1954, pp. 206–237.

Rolf Dammann: Die *Musica mathematica* von Bartolus, in *Archiv für Musikwissenchaft*, Vol. 26, 1969, pp. 140–162.

Gražina Daunoravičienė: Sound Architecture of Rytis Mažulis' Microstructural Canons, in *Menotyra* [Art Research Journal], Nr. 1, Vilnius, 2004, pp. 57–68.

Jean-Paul Davalan: Perfect Rhythmic Tilings, in *Perspectives of New Music*, Vol. 49, No. 2 (Summer), 2011, pp. 144–197.

Eric Drott: The Role of Triadic Harmony in Ligeti's Recent Music, in *Music Analysis*, Vol. 22/III, 2003, pp. 283–314.

Hans Heinrich Eggebrecht: Sinnbildlichkeit in Text und Musik bei Johann Sebastian Bach, in *Musik und Kirche*, No. 58, 1988, pp. 176–185.

Hanno Ehrler: Ordnung, Chaos und Computer. Die koreanische Komponistin Unsuk Chin, in *Musik Texte*, Vol. 96 (Februar), 2003, pp. 4–10.

Magnus Eldenius: *Formalised Composition: On the Spectral and Fractal Trails*, Doctoral Dissertation, University of Gothenburg, School of Music and Musicology, 1998.

Willem Elders: *Studien zur Symbolik in der Musik der alten Niederländer*, Bilthoven, 1968.

Willem Elders: *Symbolic scores. Studies in the Music of the Renaissance*, Leiden, New York, Köln: E. J. Brill, 1994.

Franz Carl Endres, Annemarie Schimmel: *Das Mysterium der Zahl. Zahlensymbolik im Kulturvergleich*, Köln: Eugen Diederichs Verlag, 1984.

Fritz Feldmann: Numerorum mysteria, in *Archiv für Musikwissenschaft*, Vol. 14, 1957, pp. 102–129.

Fred Fisher: Mathematic Techniques in Music History, in *Journal College Music Symposium*, Vol. 18, No. 2, 1978, pp. 24–33.

Daniel Fischlin: *In Small Proportions: A Poetics of the English Ayre, 1596–1622*, Harvard University Press, 1970.

Reinhold Friedl: Musik und Mathematik. Zwischen Astralmystik, Zufallsoperation und Chaostheorie, in *Lettre*, No. 24, 1994, pp. 70–73.

Kyle Gann: Almost Too Beautiful, in *ArtsJournal.com*, 2003 10 26.

Kyle Gann: Success in Obscurity, in *ArtsJournal.com*, 2003 10 24.

Henrik Glahn: Number symbolism and other speculative traits in two Marian motets from a 16th-century Danish manuscript, in *Analytica. Studies in honour of Ingmar Bengsson*, Stockholm, 1985, pp. 133–148.

Bernhard Haas: Veronica Diederen, Über Proportionen in der Musik des 17. und frühen 18. Jahrhunderts, in *Musiktheorie*, 17. Jahrgang, Heft 2, 2002, pp. 107–115.

Harry Hahn: *Symbol und Glaube im 1. Teil des Wohltemperierten Klaviers von Joh. Seb. Bach*, Wiesbaden: Breitkopf & Härtel, 1973.

Trudi Hammel Garland, Charity Vaughan Kahn: *Math and Music: Harmonious Connections*, Dale Seymour Publications.

Dieter Heikamp: Zur Struktur der Messe 'L'omme arme super voces musicales' von Josquin Desprez, in *Musikforschung*, Vol. 19, 1960, pp. 121–141.

Hanns-Werner Heister: Zahlenzauber und Musikmagie, in *Neue Zeitschrift für Musik*, Vol. 162 (1), 2001, pp. 10–15.

Eva Hirtler: *Die Musik als scientia mathematica von der Spätantike bis zum Barock*, Europäische Hochschulschriften. Reihe XXXVI, Musikwissenschaft, Bd. 137, Frankfurt am Main: Peter Lang, 1995.

Golden Section and Music, J. Condat (ed.), Frankfurt am Main: Peter Lang, 1988.

Gisela Gronemeyer, Reinhard Oehlschlägel: I don't Compose, I Assemble. Morton Feldman im Gespräch über *Crippled Symmetry*, in *Musik Texte*, No. 66 (November), 1996, pp. 33–36.

Joel K. Haack: The Mathematics of Steve Reich's *Clapping Music*, in *Bridges: Mathematical Connections in Art, Music, and Science. Conference Proceedings 1998*, R. Sarhangi (ed.), Winfield, Kansas: Southwestern College, 1998, pp. 87–92.

Leon Harkleroad: *The Math Behind the Music*, Cambridge University Press, 2006.

Wolfgang Hofmann: *Goldener Schnitt und Komposition. Versuch zur Fixierung eines Ordnungsprinzips*, Wilhelmshaven: Heinrichshofen, 1973.

Jadranka Hofman-Jablan: Tonality/Atonality. Order/Disorder or Order/ New Order, in *Electronic Journal Visual Mathematics*, Vol. 1, No. 1, S. Jablan & L. Radovic (eds.), Belgrade: Mathematical Institute, 1999, http://www.mi.sanu.ac.rs/vismath/jhj/index.html [accessed 12 September 2016].

Wolfgang Hufschmidt: Musik aus Zahlen. Prinzipien der musikalischen Formgestaltung in der seriellen Kompositionspraxis, in *Reflexionen über Musik heute, Texte und Analysen*, W. Gruhn (ed.), Mainz, 1981, pp. 24–57.

Tom Johnson: *Self similar melodies*, Paris: Editions 75, 1997.

Kevin Jones: Compositional Applications of Stochastic Processes, in *Computer Music Journal*, Vol. 5, No. 2 (Summer), 1981, pp. 45–61.

Herbert A. Kellner: Welches Zahlenalphabet benützte der Thomaskantor Kuhnau?, in *Die Musikforschung*, No. 33, 1980, pp. 124–125.

Herbert A. Kellner: Göttliche Unität und mathematische Ordnung. Zahlenalphabet und Gematria von A. Werckmeister bis J. S. Bach, in *Österreichische Musikzeitschrift*, Vol. 55 (11/12), 2000, pp. 8–16.

Stanley V. Kleppinger: Metrical Issues in John Adams' *Short Ride in a Fast Machine*, in *Indiana Theory Review*, Vol. 22, No. 1 (Spring), 2001, pp. 65–81.

Sebastian Klotz: Ars combinatoria oder „Musik ohne Kopfzerbrechen". Kalküle des Musikalischen von Kircher bis Kirnberger, in *Musiktheorie*, 14. Jahrgang, Heft 3, 1999, pp. 231–245.

Jeremy Leach, John Fitch: Nature, Music, and Algorithmic Composition, in *Computer Music Journal*, Vol. 19, No. 2, 1995, pp. 23–33.

Daphne Leong: Metric Conflict in the First Movement of Bartók's Sonata for Two Pianos and Percussion, in *Theory and Practice*, Vol. 24, 1999, pp. 57–90.

Wolfgang Lessing: Hommage a Sacher. Henri Dutilleux und Heinz Holligers Kompositionen für Violoncello solo über den Namen SACHER, in *Musiktheorie*, 6. Jahrgang, Heft 2, 1991, pp. 151–180.

György Ligeti: States, Events, Transformations, in *Perspectives of New Music*, Vol. 31, No. 1 (Winter), 1993, pp. 164–265.

Mark Lindley, Ronald Turner-Smith: *Mathematical Models of Musical Scales: a New Approach*, Bonn, 1993.

Marina Lobanova: Zahlen, Mystik, Magie: neueste Erkenntnisse zu Skrjabins 'Prométhée', in *Das Orchester*, Vol. 50, No. 1, 2002, 99. 6–13.

Heinz von Lösch: Eine fraktale Seepferdchen-Etüde. György Ligeti und die Chaosforschung, in *Positionen*, Heft 11, 1992, pp. 7–12.

Diane Luchese: Come un meccanismo di precisione: The Third Movement of Ligeti's Second String Quartet, in *Bridges: Mathematical Connections in Art, Music, and Science. Conference Proceedings 2001*, R. Sarhangi & S. Jablan (eds.), Winfield, Kansas: Southwestern College, 2001, pp. 37–46.

Miranda Lundy, Anthony Ashton, Jason Martineau, Daud Sutton, and John Martineau: *Quadrivium: The Four Classical Liberal Arts of Number, Geometry, Music, & Cosmology*, Wooden Books series, Walker, 2010.

Michael McNabb: Dreamsong. The Composition, in *Computer Music Journal*, Vol. 5, No. 4 (Winter), 1981, pp. 36–54.

Mensura, Mass, Zahl, Zahlensymbolik im Mittelalter, Miscellanea Mediaevalia, Veröffentlichungen des Thomas-Instituts der Universität zu Köln, A. Zimmermann (ed.), Berlin-New York: Walter de Gruyter, Bd. 16/1, 1983, Bd. 16/2, 1984.

Olivier Messiaen: *The Technique of My Musical Language*, Paris: Alphonse Leduc, 1956.

Heinz Meyer, Rudolf Suntrup: *Lexikon der mittelalterichen Zahlenbedeutungen*, München, 1987.

Ulrich Meyer: Johann Jacob Schmidts "Biblischer Mathematicus" von 1736 und seine Bedeutung für das Verständnis der Zahlensymbolik im Werk J. S. Bachs, in *Die Musikforschung*, Vol. 32, 1979, pp. 150–153.

Ulrich Meyer: Zum Problem der Zahlen in Johann Sebastian Bachs Werk, in *Musik und Kirche*, No. 49, 1979, pp. 58–71.

Ulrich Meyer: Zahlenalphabet bei J.S.Bach? – Zur antikabbalistischen Tradition im Luthertum, in *Musik und Kirche*, No. 51, 1981, pp. 15–19.

Artemis Moroni, Jônatas Manzolli, Fernando Von Zuben, and Ricardo Gudwin: Vox Populi: An Interactive Evolutionary System for Algorithmic Music Composition, in *Leonardo Music Journal*, Vol. 10, 2000, pp. 49–54.

Helga de la Motte-Haber: Alles ist Zahl. Formen pythagoreischen Denkens in der Musik, in *Musik befragt – Musik vermittelt: Peter Rummenholler zum 60. Geburtstag*, T. Ott & H. von Loesch (eds.), Augsburg: Bernd Wissner, 1996, pp. 153–163.

Helga de la Motte-Haber: Die Bedeutung von Zahlen als Grundlage musikalischer Poetik im 20. Jahrhundert, in *Magie der Zahl in der Kunst des 20. Jahrhunderts. Katalog zur gleichnamigen Ausstellung*, K. von Maur (ed.), Stuttgart, 1997, pp. 304–310.

Music and Mathematics: From Pythagoras to Fractals, J. Fauvel, R. Flood & R. J. Wilson (eds.), Oxford University Press, 2003.

Musique et nombre, Les cahiers du crem, No. 1 & 2 (Decembre), M. Biget (ed.), Centre de Recherches en Esthétique Musicale, 1986.

Gösta Neuwirth: Erzählung von Zahlen, in *Musik-Konzepte*, Bd. 26/27 'Josquin des Pres', 1982, pp. 4–38.

Catherine Nolan: Music Theory and Mathematics, in *The Cambridge History of Western Music Theory*, T. Christensen (ed.), Cambridge: Cambridge University Press, 2002, pp. 272–304.

Nombre d'Or et Musique, Condat Jean-Bernard (ed.), Quelen und Studien zur Musikgeschichte von der Antike bis in die Gegenwart, Band 19, Frankfurt am Main, Bern, New York, Paris: Peter Lang, 1988.

Hugo Norden: Per Nørgård's *Canon*, in *Fibonacci Quarterly*, Vol. 14, 1976, pp. 126–128.

Michael Nyman: *Experimental Music. Cage and Beyond*, Cambridge: University Press, 1999.

Robert W. Peck: An Overview of Mathematical Techniques in Musical Analysis, American Mathematical Society Spring Southeastern Section Conference 2003, an article submission to *Mathematics Teacher*, 23 March 2004.

[Petrov] Ю. П. Петров: Символика и диалектика чисел в «Хорошо Темперированном Клавире» И. С. Баха (I том) [Number Symbolics and Dialectics in the WTC by Bach], in *Интерпретация клавирных сочинений И. С. Баха* [Interpretation of Bach's Works for Clavier], J. P. Petrov (ed.), Moscow: Gnessin Institute, 1990, pp. 5–32.

Ludwig Prautzsch: *Vor deinen Thron tret ich Hiermit. Figuren und Symbole in den letzten Werken Johann Sebastian Bachs*, Neuhausen, Stuttgart: Hänssler, 1950.

Leonard G. Ratner: Ars combinatoria; chance and choice in eighteenth-century music, in *Studies in 18th century Music*, H. C. Robbins (ed.), London: Allen & Unwin, 1970, pp. 343–363.

Hans-Jörg Rechtsteiner: *Alles geordnet mit Maß, Zahl und Gewicht: der Idealplan von Johann Sebastian Bachs Kunst der Fuge*, Europäische Hochschulschriften, Reihe 36, Frankfurt am Main: Peter Lang, 1995.

Wolfgang Reich: Semantische und formale Gestaltungsprinzipien in den Biblischen Historien von Johann Kuhnau, in *Archiv für Musikwissenschaft*, 1958, pp. 276–290.

Peter Reichert: Tactus, Tempo und Proportion in der Orgelmusik des frühen Barock, in *Acta organologica*, Vol. 25, 1997, pp. 203–222.

Lukas Richter: Struktur und Rezeption antiker Planetenskalen, in *Die Musikforsching*, 52. Jahrgang, Heft 3, 1999, pp. 289–305.

John Rothgeb: Some Uses of Mathematical Concepts in Theories of Music, in *Journal of Music Theory*, Vol. 10, New Haven, 1966, pp. 200–215.

Jörg Rothkamm: Ordnung und Unordnung. Neues Polyrhythmik in György Ligetis Klavieretüde *Desordre*, in *Musiktheorie*, 19. Jahrgang, Heft 1, 2004, pp. 63–68.

Edward Rothstein: The Links between Maths and Music, in *The Norton Reader*, A. M. Eastman (ed.), New York, Vol. 6, 1984, pp. 1065–1070.

Frank P. Samarotto: The Body that Beats. Review of Harald Krebs *Fantasy Pieces: Metrical Dissonance in the Music of Robert Schumann*, in *Music Theory Online*, Vol. 6, No. 4 (October), 2000, http://www.mtosmt.org/issues/mto.00.6.4/mto.00.6.4.samarotto.html [accessed 12 September 2016].

Erwin Schadel: *Musik als Trinitätssymbol*, Frankfurt am Main-Berlin-Bern-New York-Paris-Wien, Lang, 1995.

Hannes Schütz: Musik und Chaostheorie zu Ligetis Klavieretude N. 1 *Désordre*, in *Musica*, Vol. 50, 1996, pp. 170–176.

Robert K. Schwarz: Steve Reich. Music as Gradual Process, in *Perspectives of New Music*, Vol. XIX, 1980–1981, pp. 373–392, Vol. XX, 1981–1982, pp. 225–286.

Julie A. Scrivener: Applications of Fractal Geometry to the Player Piano Music of Conlon Nancarrow, in *Bridges: Mathematical Connections in Art, Music, and Science. Conference Proceedings 2000*, R. Sarhangi (ed.), Winfield, Kansas: Southwestern College, 2000, pp. 185–192.

Julie A. Scrivener: The Use of Ratios in the Player Piano Studies of Conlon Nancarrow, in *Bridges: Mathematical Connections in Art, Music, and Science. Conference Proceedings 2001*, R. Sarhangi & S. Jablan (eds.), Winfield, Kansas: Southwestern College, 2001, pp. 159–166.

Dmitri N. Smirnov: *Music and Morse Code*, St. Albans, 2005, https://www.scribd.com/document/51548389/Morse-Music [accessed 12 September 2016].

Martin Supper: *Elektroakustische Musik und Computermusik*, Darmstadt, 1997.

Wolfgang Martin Stroh: Mathematik und Musikterminologie, in *Zur Terminologie der Musik des 20. Jh.*, H. H. Eggebrecht (ed.), Stutgart, 1974, pp. 33–59.

The OM Composer's Book, Vol. 1, coll. Musique/Sciences, A. Agon, G. Assayag & J. Bresson (eds.), Paris: Editions Delatour France, Ircam-Centre Pompidou, 2006.

The OM Composer's Book, Vol. 2, coll. Musique/Sciences, J. Bresson, A. Agon & G. Assayag (eds.), Paris: Editions Delatour France, Ircam-Centre Pompidou, 2008.

[Tsaregradskaya] Татьяна В. Цареградская: *Время и ритм в творчестве Оливье Мессиана* [Time and Rhythm in the Work of Olivier Messiaen], Moscow: Klassika, 2002.

Annette Veit: *Zur Bedeutung der Zahl in der Musik*, Diplomarbeit am Institut für Musikanalytik, Wien, 1991.

Martin Vogel: *Die Zahl 7 in der spekulativen Musik Theorie*, Inaugural Dissertation zur Erlangung der Doktorwürde Genehmigt von der Philosophischen Fakultät, Bonn, 1955.

Johann Gottfried Walther: *Praecepta der musikalischen Composition*, Weimar, 1708/Leipzig: Breitkopf & Härtel, 1955.

Eric Werner: The Last Pythagorean Musician: Johannes Kepler, in *Aspects of Medieval and Renaissance Music*, New York, 1966, pp. 867–882.

Rudolf Wille: Mathematische Sprache in der Musiktheorie, in *Jahrbuch Überblicke Mathematik*, Manheim: Bibliographisches Institut, 1980, pp. 167–184.

Peter N. Wilson: Formalisierte Musik – naturwissenschaftliches Denken in der Kunst. Gedanken zu den stochastischen Kompositionstheorien von Iannis Xenakis und Clarence Barlow, in *Neuland*, Vol. 5, 1985, pp. 52–59.

David Wright: *Mathematics and Music*, Mathematical World, Vol. 28, Providence, RI: American Mathematical Society, 2009.

Zeichen und Struktur in der Musik der Renaissance, Hortschansky Klaus (ed.), Kassel, Basel, London, New York: Bärenreiter, 1989.

Index of Names

About the Author

RIMA POVILIONIENĖ (born 1975), musicologist, PhD. She holds a senior researcher position at the International Semiotics Institute (ISI) at Kaunas University of Technology. Rima is an associate professor in the Department of Musicology of the Lithuanian Academy of Music and Theatre, the assistant editor-in-chief of the scientific yearly *Lietuvos muzikologija* [Lithuanian Musicology] and an editor at the Lithuanian National Philharmonic. She was on internships at the Institute of Musicology at Leipzig University (2004) and IRCAM (2012). She has edited over 10 collections and published more than 150 critic reviews. Her monograph *Musica Mathematica* (in Lithuanian, 2013) was awarded the Vytautas Landsbergis prize for the best musicological work.

List of author's scientific studies and publications on the subject of *musica mathematica*:

Prasminis ir konstruktyvinis skaičiaus principas baroko garsų mene [Numerical Semantics and Construction in Baroque Music], Bachelor's thesis, Vilnius: Lithuanian Academy of Music and Theatre, 2001.

Prasminis ir konstruktyvinis skaitmens principas baroko garsų mene [Semantic and Constructive Meaning of Number in Baroque Music], in *Lietuvos muzikologija* [Lithuanian Musicology Journal], No. 3, Vilnius, 2002, pp. 17–40. ISSN 1392–9313.

Konstruktyvinės ir simbolinės skaičiaus funkcijos XX a. muzikoje [The Role of Numerical Construction and Symbols in 20th Century Music], Master's thesis, Vilnius: Lithuanian Academy of Music and Theatre, 2003.

Konstruktyvinės ir simbolinės skaitmens funkcijos XX a. muzikoje [Constructive and Symbolical Functions of Number in the 20th Century Music], in *Lietuvos muzikologija* [Lithuanian Musicology Journal], No. 5, Vilnius, 2004, pp. 61–83. ISSN 1392–9313.

Muzika – tai skambantys skaičiai [Music as Sounding Numbers], in *Kultūros barai* [Culture Journal], No. 3 (472), Vilnius, 2004, pp. 31–34. ISSN 0134–3106.

Technologinis ir simbolinis skaitmens konstruktyvizmas George'o Crumbo *Black Angels* ir György Ligeti *Etudes* [Technical and Symbolic Numerical Construction in George Crumb's *Black Angels* and György Ligeti's *Etudes*], in *Tiltai* [Bridges Journal], No. 21, Klaipėda, 2004, pp. 137–148. ISSN 1648–3979.

Baroko muzikos numerologija [Baroque Musical Numerology], in *Muzikos kalba* [Musical Language], Vol. 2, Vilnius: Enciklopedija, 2006, pp. 508–564 (co-author with Gražina Daunoravičienė). ISBN 998643338X.

Music Composition as a Numerical Construction. The Projections of Magic Number Quadrates and Dimensions of Ziggurat Pyramids in the Composition *Ziqquratu* (1998) by Šarūnas Nakas, in *Musical Work: Boundaries and Interpretations*, A. Žiūraitytė (ed.), Vilnius, 2006, pp. 278–291. ISBN 9986–9069-5-4.

Perceiving and Cognizing the Mathematical Processes in Music Composition in 20th Century, in *ICMPC9. Proceedings of the 9th ICMPC,* Bologna: University of Bologna, 2006, pp. 932–937. ISBN 88–7395-155-4.

Several Mathematical Aspects of Music Notation in the 20th Century Music Composition, in *Principles of Music Composing*, No. 7, Vilnius, 2007, pp. 60–66. ISBN 978–9986-503-71-2.

Numerologinių technikų inovacijos XX a. antrosios pusės muzikos kompozicijoje [Innovative Digital Technologies in the Musical Composition of the Second Half of the 20th Century], in *Lietuvos muzikologija* [Lithuanian Musicology Journal], No. 9, Vilnius, 2008, pp. 70–89. ISSN 1392–9313.

Antikiniai *mathesis* kaip garsų meno grožio aspektai [Aspects of Ancient Mathesis as the Beauty of Sound Art], in *Menotyra* [Art Research Journal], Vol. 16, No. 1–2, Vilnius, 2009, pp. 1–11. ISSN 1392–1002.

Fugue in B-flat Minor by Čiurlionis: Structural and Semantic Analysis, in *Principles of Music Composing,* No. 11, Vilnius, 2011, pp. 23–30. ISBN 978–609-8038-03-3, ISBN 978–609-8071-05-4.

Muzikinio ir verbalinio menų sąveikos fenomenas: kriptografinės muzikos komponavimo praktikos aspektai [Phenomenon of Interaction of Musical and Verbal Arts: Aspects of Cryptographic Practice in Music Composition], in *Lietuvos muzikologija* [Lithuanian Musicology Journal], No. 12, Vilnius, 2011, pp. 34–46. ISSN 1392–9313.

Čiurlionio autografo ir dedikacijos tyrimai. Fuga *in b*: struktūrinė ir semantinė analizė [Search of Čiurlionis Signature and Dedication. Fugue in B-flat: Structural and Semantic Analysis], in *Mikalojus Konstantinas Čiurlionis (1875–1911): jo laikas ir mūsų laikas* [Mikalojus Konstantinas Čiurlionis (1875–1911): His Time and Our Time], G. Daunoravičienė & R. Povilionienė (eds.), Vilnius: Lithuanian Academy of Music and Theatre, 2013, pp. 268–301. ISBN 978–609-8071-11-5.

Frozen Music – Dialogues between Music and Architecture, in *Music and Technologies*, D. Kučinskas (ed.), Cambridge Scholars Publishing, 2013, pp. 83–98. ISBN (10): 1–4438-4213-3, ISBN (13): 978–1-4438-4213-6.

Musica mathematica. Tradicijos ir inovacijos šiuolaikinėje muzikoje [Musica mathematica. Traditions and Innovations in Contemporary Music]. Vilnius: Lithuanian Academy of Music and Theatre, 2013, 256 p. ISBN 978–609-8071-10-8.

Interaction of Constructivism and Semantics in Bronius Kutavičius' Musical Language, in *Music That Changed Time*, R. Povilionienė & J. Katinaitė (eds.), Vilnius: Lithuanian Composers' Union, Lithuanian Academy of Music and Theatre, 2014, pp. 155–163. ISBN 978–609-8038-07-1, ISBN 978–609-8071-21-4.

Methodology of Music Research
Methodologie der Musikforschung

Edited by / Herausgegeben von Nico Schüler

Vol. 1 Nico Schüler (Ed.): Computer-Applications in Music Research. Concepts, Methods, Results. 2002.

Vol. 2 Mirjana Veselinović-Hofman: Fragmente zur musikalischen Postmoderne. 2003.

Vol. 3 Otto E. Laske: Musikalische Grammatik und Musikalisches Problemlösen. Utrechter Schriften (1970–1974). Herausgegeben von Nico Schüler. 2004.

Vol. 4 Nico Schüler (Ed.): On Methods of Music Theory and (Ethno-) Musicology. From Interdisciplinary Research to Teaching. 2005.

Vol. 5 Leon Stefanija / Nico Schüler / Tuomas Eerola / Reiko Graham / Vanessa Nering / Mirjana Veselinović-Hofman: Musical Listening Habits of College Students in Finland, Slovenia, South Africa, and Texas. Similarities and Differences. 2010.

Vol. 6 Leon Stefanija / Nico Schüler (Eds.): Approaches to Music Research. Between Practice and Epistemology. 2011.

Vol. 7 Denis Collins (Ed.): Music Theory and its Methods. Structures, Challenges, Directions. 2013.

Vol. 8 Nico Schüler: Computer-Assisted Music Analysis (1950s-1970s). Essays and Bibliographies. 2014.

Vol. 9 Rima Povilionienė: Musica Mathematica. Traditions and Innovations in Contemporary Music. 2016.

www.peterlang.com

www.ingramcontent.com/pod-product-compliance
Ingram Content Group UK Ltd.
Pitfield, Milton Keynes, MK11 3LW, UK
UKHW020835290926
15470UKWH00004B/1352

9 783631 713815